SpringerBriefs in Ethics

More information about this series at http://www.springer.com/series/10184

Paul Walker · Terence Lovat

Life and Death Decisions in the Clinical Setting

Moral Decision Making Through Dialogic Consensus

 Springer

Paul Walker
School of Medicine and Public Health
The University of Newcastle
Newcastle, NSW
Australia

Terence Lovat
School of Humanities and Social Science
The University of Newcastle
Newcastle, NSW
Australia

and

University of Oxford
Oxford
UK

ISSN 2211-8101   ISSN 2211-811X   (electronic)
SpringerBriefs in Ethics
ISBN 978-981-10-4300-0   ISBN 978-981-10-4301-7   (eBook)
DOI 10.1007/978-981-10-4301-7

Library of Congress Control Number: 2017934630

Printed on acid-free paper

This Springer imprint is published by Springer Nature
The registered company is Springer Nature Singapore Pte Ltd.
The registered company address is: 152 Beach Road, #21-01/04 Gateway East, Singapore 189721, Singapore

Kenneth Edward Walker, RIP 7th August 2015

Preface

Moral decision making in clinical settings, especially around life and death decisions, has never been easy. In our current context, we see life-prolonging technological advancements racing ahead of our reflection on when to employ those advancements. Hence, decision making which aims to truly maximise the good of the patient has become ever more complex. At the same time, societies themselves have become more complex, with the largely homogenous societies of the past giving way to increasingly multicultural, multifaith ones. Hence, the relatively predictable set of values that once might have defined the stakeholders involved in difficult moral decision making has been replaced by value pluralism. As a result, the lines of Western philosophical thought that have determined such decision making in the past need to be reappraised and recalibrated to take account of this new situation. These lines of thought have resulted in the *substantive* (that is, stand-alone) ethical frameworks of deontology (also known as categorical or intrinsic), teleology (consequentialism, utilitarianism), and virtue ethics. Traditionally, one or other or a combination of these three frameworks has guided ethical decision making in the Western clinical setting.

Deontology argues that some things must not be done. For example, any act that could be described as killing is impermissible because of the intrinsic nature of the act of killing. Teleology argues that what is permissibly done depends upon the consequences of the action. For example, killing might be permissible if it can be shown to be of greater benefit (in terms, for example, of relieving suffering) than not killing. Virtue ethics focus on the character of the person who is doing the act, rather than the act itself; in the clinical setting, moral decision making turns on maximising the good of the patient through empathic, compassionate caring.

Four principles distilled from these frameworks—autonomy, non-maleficence, beneficence and justice—have historically guided ethical decision making in clinical settings. In the context of value pluralism, however, recourse to these frameworks alone has potential to overlook the essential inter-connectedness within the community, which is necessary for moral decision making. Regardless of how one's own ethical values, conceptions of the good or life-choices might be reached, how do they differ from those of others, especially the stakeholders to any life and death decision? Herein, we propose a *proportionist* approach as a way of balancing out the wisdom to be found in the substantive frameworks and principles above with the realities of the new context of advanced technological potential and value pluralism.

In order to put into practice an approach of proportionate balancing of rules and consequences, a moral decision-making *process* should be followed. This process involves having a conversation, a dialogue or a discourse, with collaboration, amongst all the stakeholders. The aim of the dialogue is to reach consensus in the decision, via mutual understanding of the values held by the patient and the patient's family and others whom they see as significant, set against the concrete reality of the situation at hand. From a virtue ethics' perspective, this process seeks to maximise the various Goods of the patient so as to actualise optimal care for the patient. This process of *dialogic consensus* is inspired in part from the writings of Jürgen Habermas, a continental philosopher, political scientist and sociologist. His concepts of discourse theory of morality and principles of communicative action together underpin a moral decision-making process of inclusive, non-coercive and reflective dialogue.

The central argument of this book is that in our contemporary era, characterised medically by an ever-increasing armamentarium of life-sustaining technology, an active process of moral decision making in clinical situations, rather than mere monological contemplation on the part of a clinician, is required. Reaching unforced consensus amongst the stakeholders of clinician, patient, family and relevant others renders the decision with what is known in moral philosophy as normative force. Normative force means that the decision has a sense of *oughtness* or *shouldness* associated with it. Additionally, if this process of non-coercive dialogic consensus is understood and reflected upon, the patient, family and others are less likely to have lingering doubts about whether the normatively right decision is being made.

Newcastle, Australia   Paul Walker
Newcastle, Australia/Oxford, UK   Terence Lovat

Contents

Chapter 1
Introduction: The Current Dilemma and the Need for Moral Philosophy

1.1 Why Does Medicine Need Moral Philosophy?

Certain actions and behaviours are recognised by individuals and society as 'Good', while others are recognised as 'Bad', and as such, they constitute the borders which delimit what are referred to as moral standards. These standards are described as *normative* when they are associated with a sense of *oughtness* or *shouldness* (notions which are further discussed below). Collectively, these standards comprise morality—the summation of the value systems that guide our actions and behaviours as human beings. That is to say, when we are faced with a decision, except in completely banal matters such as whether to have milk in coffee or not, each of us recognise at some level, that there is a better choice and a worse choice that we could make (Dworkin 2011). In a holistic sense, morality concerns *all* that is significant, or that matters to humans.

The aim of moral philosophy is to find a way of thinking better about moral questions (Hare 1989), the study of how we ought to act (Singer 1994), the teaching of critical reasoning, and the enquiry into what is good (Moore 1903, Chap. 1, Sect. 2). This enquiry is informed by the question "how should I act?"; but should also look to our intentions and our motivations. Thus, we view, as a more apposite formulation, the question of Socrates "how should I live?" (Plato 1997, [352d]).

There was a time when philosophers were the leaders in asking the big questions about life: What is a good life? How should we live? Where do we go after death? How do we make a society more just? Indeed, in Greco-Roman times, philosophers enjoyed greater prestige than physicians. As the United Nations Education and Scientific Organisation (UNESCO) notes, philosophy implies freedom, in and through reflection—'because it is a matter not just of knowing, but of understanding' (Matsuura 2007). As doctors, and as people, we (most of us anyway) want to do good, to be good, to seek wisdom. Wisdom, in the sense of understanding, rather than merely knowing, segues into values, and hence morality, and is integral to decision making in clinical contexts.

© The Author(s) 2017
P. Walker and T. Lovat, *Life and Death Decisions in the Clinical Setting*,
SpringerBriefs in Ethics, DOI 10.1007/978-981-10-4301-7_1

The wisdom we seek in making moral decisions is to understand what we ought to strive after, what we should do. Hence, in everyday usage, the language of morality is often characterised by words such as *ought* and *should*. These words are important to conceptions of moral behaviour, and hence are referred to by moral philosophers as 'decisive-reason-implying concepts' (Parfit 2011). First, they assert an obligation (Ross 2002). That is, there is a logical inconsistency if individual A ought to do something, but individual B, in the same situation, is not expected to respond to the 'ought' and do the same thing (Hare 1993). Hence, the action which ought to be taken can be generalised to others in precisely the same situation. Second, it is expected that its use can be justified—an action *ought* to be done— because of a reason that is able to be articulated, or because, in that situation, acting in a certain way is viewed as right, just, good, or truthful. According to the understanding to be found in this book, *normative* questions motivate action – that is, they make *claims* upon us; they command, oblige, recommend, or guide (Korsgaard 1996). It is in this sense that medicine needs moral philosophy. Without it, the wisdom to know what to do, how to behave and what standards to apply to moral decision making, would be absent.

1.2 Ethical Conflicts

A moral conflict exists when, faced with a decision-making situation, two (or more) actions are available, each of which appears to have an element of permissibility, or moral *oughtness* to it. A moral dilemma exists when a person's reasoning appears to require that she do each of the two (or more) actions, which might even be opposing or mutually exclusive, at the same time.

One example of an ethical dilemma is that of an impoverished father who steals food only to feed his starving family. He both ought not to steal and he ought to prevent his family from starving to death. A second example is that a borrowed weapon should not be returned to an owner when he is not in his right senses, despite the existing duty to repay a debt (Plato 1952). Ordering or ranking of 'oughts' to constitute a hierarchy could be a helpful action in these sorts of situations. A third example, in a clinical setting, is when a clinician who does not believe in deliberately shortening a patient's life considers prescribing an opioid aimed at the relief of pain and suffering but which will likely also shorten the patient's life. This particular dilemma is explored below in Sect. 2.3. Another sort of ethical dilemma arises when someone is in a situation where two (or more) opposing and mutually exclusive actions are available, but the evidence for each is incomplete or inconclusive. It is not uncommon in medicine that there is a degree of uncertainty associated with the outcome of different therapeutic options.

It is the understanding behind this book that moral conflicts and dilemmas are inevitable. As will be explained, this is at least partly because of the pluralistic and fragmented character of contemporary society. Dialogue or discourse seeks to clarify the ethical imperatives in the situation at hand, as each participant perceives them. This dialogue lays the groundwork for the resolution of differences that persist.

1.3 Ethics or Morals?

Mindful of this plural and fragmented character of contemporary society, and thus the need to clarify the different perspectives of those in the dialogue, it is important at this early stage, to consider whether there is any difference between the words *ethics* and *morals*, and whether that difference is important (Walker and Lovat 2017).

The first international conference devoted to medical ethics, in France in 1955, used the term *medical morality*. Subsequent to this, "bioethics" was a term newly coined in North America in the 1960s to refer to 'the rise of professional and public interest in moral, social, and religious issues connected with the "new biology" and medicine' (Fox and Swazey 1984). It is clear that clinicians often face moral dilemmas when caring for patients—'[t]o help them cope with these, biomedical ethics aims to implement moral norms for particular problems and contexts' (Tassy et al. 2007). Available sources which clinicians can make appeal to in order to guide their ethical decisions include: the codes of conduct or ethical statements of their professional bodies (medical associations, specialist associations, or nursing associations); world ethics declarations; white papers; or perhaps prior decisions of clinical ethics committees, amongst others. These may however only set limits to what is ethically appropriate or inappropriate, and therefore might lack applicability to individual dilemmas that clinicians and their patients face together. H. Tristram Engelhardt writes that the answer to the intellectual question "How can I consistently understand what is right conduct in the health care professions … and justify it to others"? is to be found through a philosophical approach rather than empirical anthropology or sociology (Engelhardt 1996).

The etymological origins of the words *ethics* and of *morals* derive, respectively, from the words in Greek (*êthos, ethos*) referring to 'character', and Latin (*mos, mores, moralis*) referring to 'customs' (Cassin et al. 2014). However, in contemporary usage, the words are employed interchangeably. Consideration of their respective Greek and Latin origins and uses is ambiguous. This is partly because Cicero chose to translate Aristotle's Greek *êthos* (from which our word *ethics* derives) into Latin as *moralis* (from which our word *morals* derives). From another perspective, in terms of their use by Greek and Latin writers, morality has a prescriptive dimension. Hence, morality may also be conceived of as applied ethics—putting the theory of ethics into actual practice.

Etymological ambiguities aside, important differences are consistently able to be recognised in the philosophical literature. Here, *ethics* refers to 'character, personal disposition' (Onions 1966), is 'a matter of personal worldview'(Ormerod and Ulrich 2013), 'come from within … a personal sense of right and wrong' (O'Conner and Stewart 2012), and within which is found 'forms of life according to which individuals assess the good life for themselves' (Forst 2014). On the other hand, *morals* refers to 'rules or principles governing human behaviour which apply universally within a community' (Strawson 1961), emphasises 'more the sense of social expectation' (Williams 2006), the 'shared public or communal norms about right and wrong actions' (Kerridge et al. 2013), 'what is acceptable and right to all the parties concerned' (Ormerod and Ulrich 2013), and 'the complex of norms, behavioral models, virtues, and values which characterize society' (Besio and Pronzini 2014). This book emphasises the understanding that ethics refers to a more individual or organisational assessment of values as relatively good or relatively bad, while morals connotes a more collective and intersubjective assessment of what is Right or Just for all affected. Put another way, ethics makes claims about how people should act in their own lives, in order to live a good life from the perspective of their own values. Whereas morals make claims about how people should all act together, in order that they all live a good life, allowing for their different but necessarily co-existing value perspectives.

Jürgen Habermas is a member of the second generation of the Frankfurt school (the Institute for Social Research, Goethe University Frankfurt). His expertise straddles protean disciplines, but especially sociology, political science and philosophy. He has argued that morality is a protective safety device for people, both as individuals and collectively, who co-exist through inter-dependency within society. He characterises this as a 'densely woven fabric of mutual … exposedness and vulnerability' (Habermas 1990). According to this understanding, the vulnerability of human beings requires a system of mutual consideration—defending the integrity of the individual but also the collective. He refers to suicide, for example, as both an individual failing but moreover the collective failing of an inter-subjectively shared lifeworld. The systems of morality must simultaneously emphasize the inviolability of the individual by postulating equal respect for the dignity of each individual. At the same time, however, systems of morality must also protect the web of intersubjective relations by which a community survives. These principles correspond to justice, coequal respect and rights and to what Habermas refers to as solidarity, but which others may refer to as beneficence or compassion, empathy or care for our neighbours.

Hence, the language of *ethics* and *morals* both have a place in the scheme of thinking in this book, but they are more than interchangeable terms. They connote, respectively, the more individualistic (ethics) and communal (morals) dimensions of the *oughtness* of standards of right and wrong, good and bad, or better and worse decision-making choices.

1.4 The Rationale for This Book

We offer three reasons for this book.

First, we believe that since each clinical doctor-patient contact inherently involves dealing with another human being, the clinical doctor-patient relationship has a necessary foundation in moral philosophy. We believe that the clinical encounter is grounded in inter-connectedness amongst the clinician, the patient, and the patient's family and wider community. Clinical interactions are, properly speaking, moral decision-making situations. This applies to all medical consultations, not just complex and challenging decisions in life and death situations. Our aim should be to maximise the good of the patient. Achieving this requires discussion and consensus-seeking. We argue that the historical emphasis of medical ethics, based on substantive frameworks and principles derived from them, is no longer sufficient, least of all in our contemporary era. We borrow from the philosophical stream known as *phenomenology* to encompass, within the clinical encounter, awareness of others, recognition of patients' actual reality and our encapsulation within that reality—that is within the world of lived experience. According to this understanding of phenomenology, the clinical encounter takes as its starting point the seeing of patients' ethical dilemmas as being grounded in their concrete existential situations (Carel 2011).

Second, our contemporary era is characterised by disparate value pluralism, and hence ethical and moral pluralism. That is, widespread immigration and global communication results in people holding a wide range of beliefs and values that are important to them. Our communities are increasingly characterised by socio-cultural, ethno-national and religious diversity, which potentially results in conflicting life views.

Third, we distinguish ethical decision making from moral decision making. We see ethical decision making as a relativistic and essentially subjective monologue determined by one's personal value system. On the other hand, we see moral decision making, most properly, as a communal *process*, a process of dialogue amongst those most affected by the decision—in this case, amongst clinician, patient, family and their significant others. This dialogue might be characterised as being inclusive, non-coercive and self-reflective. We argue that this approach is most apposite to moral decision making in clinical situations, especially given the plurality of our contemporary society.

A 2005 report from a panel of bioethicists that sought to determine the top ten ethical challenges facing the public identified the top challenge as 'disagreement between patients/families and health care professionals about treatment decisions' (Breslin et al. 2005). Hence, the decision-making process that we argue in favour of here might also be useful to non-medically qualified health care providers and educators who provide clinical face-to-face care for patients. These would include senior nursing staff and allied health workers acting relatively autonomously, as well as psychologists, social workers and chaplains. We envisage that the approach of a *process* for moral decision making in clinical situations which, by the end of this book, we will term *dialogic consensus,* is, or will become integral in multi-disciplinary case conferences and clinical ethics consultations, as well as in individual consultations amongst clinicians and their patients.

In order to make this point more clearly: although, in this book, our examples place clear emphasis on moral decision making in clinically critical situations, we do not believe that our medical ethics toolbox should be brought out only in critical situations. Rather, we argue that since each clinician-patient contact inherently involves dealing with another human being, all clinician-patient encounters have a necessary foundation in moral philosophy. Thus, all clinical interactions are, properly speaking, moral decision-making situations.

1.5   Chapter Outline

This book is presented in six Chapters. This first, introductory, chapter introduces the book. It considers the rationale for such a book as this. It considers words including *ethics* and *morals, conflict* and *dilemma*, why there are difficulties with ethical and moral decision making in medical situations, and why this prompts a re-evaluation of moral decision making in medical situations. It introduces the premise that decision making in the clinical encounter should be approached from the perspective that the decision being made follows a process of moral discourse by way of an inclusive and non-coercive reflective dialogue which seeks consensus, as distinct from what might be termed an ethical monologue.

Chapter 2, begins with an exploration of ethical decision making through the Classical period, the Medieval period, and the Modern period of our history. It then considers the workings of three normative frameworks in the secular Western tradition that, historically, have been important in guiding decisions in medical ethics. These are deontology, teleology, and virtue ethics. The deontological framework predicates moral permissibility on the intrinsic nature of the Act. The teleological framework predicates moral permissibility on the consequences of the Act. The virtue ethical framework focuses on the character of the agent. They are substantive—by which we mean that they are stand-alone frameworks. In considering deontological principles, specific attention is given to philosophical principles which help determine who actually is a person from a moral decision-making perspective, and how we might make moral decisions at the end-of-life, for example in Intensive Care Units (ICUs). In considering teleological principles, aspects of triage are discussed. Under the virtue ethics framework, the Good of the patient is the ultimate purpose of medicine, and empathy, compassion and care are proposed to guide normative moral decision-making in clinical situations. Finally, the theistic framework of the Islamic-Judaeo-Christian tradition is considered.

Chapter 3, considers the four principles that have been distilled from the normative frameworks identified in the previous chapter, proposed as ways of guiding practical decision making in medical ethics. These are *respect for autonomy, non-maleficence, beneficence,* and *justice* (Beauchamp and Childress 2013). They have had considerable influence on moral decision making in clinical situations, and, at a minimum, they offer a common ethical language amongst clinicians. Shortcomings in their theoretical and practical application will be identified. Various understandings of autonomy—considered to be first amongst the four principles—will be considered. Critical re-examination suggests that our traditional understanding of autonomy is impoverished and requires re-evaluation. What is formed herein as the *proportionist* approach seeks a virtuous mean or balance-point in moral decision making that takes account of the frameworks and principles identified in this and the previous chapter but in a way that is more grounded in the realities of the modern era (as will be expanded upon in the following chapters). It seeks to balance intrinsic rules and empirical consequences, hence utilizing but also going beyond the bounds of the deontological and teleological frameworks alone. Its starting point is the actual reality of the patient in their situation. It is put into clinical practice via a process we term *dialogic consensus*, a term we will explore in Chaps. 4 and 5.

Chapter 4, considers what makes our contemporary era, herein termed post-modern, different from earlier eras, and why we therefore need to move from appeal to a substantive ethical framework, to an active process of moral decision making. Thus, the move is from *ego* to *alterity* (otherness) in the notion of dialogic consensus. Dialogic consensus is derived from Jürgen Habermas' notions of discourse theory of morality and communicative action. Recognition of our inter-connectedness is important for Habermas because of its contribution to normativity, in that it serves as a motivator to act, consequent upon a sense of oughtness or shouldness. His discourse theory of morality requires that the consequences for all persons affected must be considered, while his principles of communicative action imply that the discourse is based upon consensus, subsequent to inclusive, non-coercive and reflective dialogue. Intersubjective consensus after dialogue within the relevant community imbues the decision with normative force that, in turn, renders the process one which is action-guiding. Habermas' discourse theory of morality generalises and expands the Kantian categorical imperative, as determined by ethical monologue, to a wider consensus-seeking dialogue. Thus, consensual agreement is reached about what constitutes morally-correct action. Relocating decision making from a monological space, into one characterised by dialogue within the stakeholder community, is especially appropriate to the clinical encounter. This form of moral decision making is at the heart of the notion of dialogic consensus.

Chapter 5, explores some practical difficulties that will inevitably be encountered in the clinical setting characterized by expanded technological capacity and value pluralism within the stakeholder group. Responding to these entails more intensive analysis of the Habermasian notions of discourse theory of morality and communicative action, and especially aligning them with his theory of the ways of knowing. By basing discourse around a way of knowing characterized by self-reflectivity, each member of the stakeholder group can be: (1) facilitated in seeing technological capacity as merely a factor to be considered—but not necessarily the determinative factor; and, (2) to take better account of the interpretations offered by other members of the decision-making group. That is, without listening, there can be no knowing. It makes sense that we cannot have a discussion about morals if we do not have the words for the concepts and if we do not agree about the meanings of the words. We will argue that, practical difficulties in achieving the ideal dialogue notwithstanding, the process described herein has both applicability and great merit for moral decision making in clinical settings.

Chapter 6, concludes and summarises the concept of dialogic consensus presented in this book. Clinical encounters are understood as inter-relationships among persons. The final purpose of the clinical encounter is to maximise the good of the patient in all its connotations, via the provision of empathic compassionate care. As has been emphasized throughout the book, the current era is characterised by pronounced value pluralism. The proportionist approach seeks to maximise the patient's good, based on a balance between a priori imperatives and empirical utility. It begins from the reality of the patient in the situation of their illness. The practical application of the proportionist approach in clinical practice requires application of the principles contained in Habermas' discourse theory of morality and his principles of communicative action, moderated further by his ways of knowing theory. Thus, a cooperative search for truth, in order to make a properly shared decision, is undertaken. This process, named as dialogic consensus, is characterised as an inclusive, non-coercive and reflective dialogue aiming to seek consensus in the decision being made. It will be further argued that such a process has potential to address current realities in a way that is more sustainable and appropriate to the circumstances of the contemporary clinical setting.

References

Beauchamp, T.L., and J.F. Childress. 2013. *Principles of biomedical ethics*, 7th ed. New York, Oxford: Oxford University Press.
Besio, C., and A. Pronzini. 2014. Morality, ethics, and values outside and inside organisations: An example of the discourse on climate change. *Journal of Business Ethics* 119 (3): 287–300. doi:10.1007/210551-013-1641-2.
Breslin, J.M., S.K. MacRae, J. Bell, P.A. Singer, and G. University of Toronto Joint Centre for Bioethics Clinical Ethics. 2005. Top 10 health care ethics challenges facing the public: Views of Toronto bioethicists. *BMC Medical Ethics* 6: E5. doi:10.1186/1472-6939-6-5.
Carel, H. 2011. Phenomenology and its application in medicine. *Theoretical Medical Bioethics* 32 (1): 33–46. doi:10.1007/s11017-010-9161-x.
Cassin, B., M. Crepon, and F. Prost. 2014. *Morals/Ethics* (S. Rendall, C. Hubert, J. Mehlman, N. Stein, and M. Syrotinski, Trans., *Dictionary of untranslatables: A philosophical lexicon*). Princeton: Princeton University Press.
Dworkin, R. 2011. *Justice for hedgehogs*, 1st ed. Cambridge, MA: The Belknap Press of Harvard University.
Engelhardt, H.T. 1996. *The foundations of bioethics*, 2nd ed. New York: Oxford University Press.
Forst, R. 2014. *Ethics and morals* (J. Flynn, Trans., *The right to justification: Elements of a constructive theory of justice*). New York: Columbia University Press.
Fox, R.C., and J.P. Swazey. 1984. Medical morality is not bioethics—Medical ethics in China and in the United States. *Perspectives in Biology and Medicine* 27: 336–360.
Habermas, J. 1990. *Moral consciousness and communicative action* (C.L.S. Weber, Trans.). Cambridge: Polity Press.
Hare, R.M. 1989. *Essays in ethical theory*. Oxford: Clarendon Press.
Hare, R.M. 1993. Universal prescriptivism. In *A companion to ethics. Blackwell companions to philosophy*, vol. 2, ed. P. Singer, 451–463. Malden, MA: Blackwell.
Kerridge, I., M. Lowe, and C. Stewart. 2013. *Ethics and law for the health professions*, 4th ed. Sydney: The Federation Press.

Korsgaard, C.M. 1996. The normative question. In *The sources of normativity*, ed. O. O'Neill, 7–48. Cambridge: Cambridge University Press.

Matsuura, K. 2007. *Philosophy: A school of freedom* (*Teaching philosophy and learning to philosophize. Status and prospects*). In Moufida Goucha, Feriel Ait-ouyahia, Arnaud Drouet, and K. Balalovska, 303. Paris: United Nations Educational, Scientific and Cultural Organisation.

Moore, G.E. 1903. *Principia ethica*. Prometheus Books.

O'Conner, P.T., and K. Stewart. 2012. The grammarphobia blog. http://www.grammarphobia.com/blog/2012/02/ethics-vs-morals.html. Accessed January 2, 2015.

Onions, C.T. 1966. *The Oxford dictionary of English etymology*, xvi, 1024. Oxford: Clarendon.

Ormerod, R., and W. Ulrich. 2013. Operational research and ethics: A literature review. *European Journal of Operational Research* 228 (2): 291–307.

Parfit, D. 2011. *On what matters*. Oxford: Oxford University Press.

Plato. 1952. The Republic (B. Jowett, Trans.). In *The dialogues of Plato, the seventh letter. Great Books of the Western World*, vol. 7, 1st ed, 295–441. Chicago: William Benton, Encyclopaedia Britannica Inc.

Plato. 1997. Republic (G. Grube, Trans.). In *Plato: Complete works*, ed. J.M. Cooper, 971–1223. Indianapolis, Cambridge: Hackett Publishing Company.

Ross, D. 2002. In *The right and the good*, 2nd ed, ed. P. Stratton-Lake, 220. Oxford: Oxford University Press.

Singer, P. 1994. Introduction. In *Ethics. Oxford readers*, ed. P. Singer, 3–13. Oxford: Oxford University Press.

Strawson, P.F. 1961. Social morality and individual ideal. *Philosophy* 36 (136): 1–17.

Tassy, S., P. Le Coz, and B. Wicker. 2007. Current knowledge in moral cognition can improve medical ethics. *Journal of Medical Ethics* 34: 679–682.

Walker, P., and T. Lovat. 2017. Should we be talking about *ethics* or about *morals*. *Ethics and Behavior*. doi:10.1080/10508422.2016.1275968.

Williams, B. 2006. *Ethics and the limits of philosophy*. London: Routledge Classics.

Chapter 2
Traditional Approaches to Ethical Decision Making

2.1 What Can History Tell Us?

The 3000 page, 5 volume *Encyclopedia of Bioethics* notes that 'bioethics as a discipline cannot expect intellectual respect, much less legitimately affect moral behaviour, unless it can show itself to be grounded in solid theory justifying its proposed values, principles, and rules' (Post 2004). Hence, we begin with an overview of that 'solid theory'.

The ethical and moral judgements we make are necessarily situated in an historical context. A system of morals may be codified in different ways, in different cultures, and at different times. Homosexuality is an obvious historical example. Historically, in the Western tradition, it is possible to apportion the development of moral philosophy into four epochs. In each of these, appeal was made to different ethical systems or frameworks.

In the *Classical* (or *Ancient*) period are located the societies clustered around the Mediterranean—the polytheistic Hellenistic, and, to a lesser extent, Roman traditions, especially of Plato, the 'Sophists', and Aristotle. This period ended between the third and sixth century, after the sack of Rome in 410 CE, the closure of Plato's Academy in 529, and the plague which began in 542. In this era, appeal for ethical guidance was made to either the *polis* (the Greek city-state) or to one or more of their several gods.

In the *Medieval* (or) *Middle* period, classical ideas were re-cast in the light of monotheistic traditions of Augustine, Maimonides, and Aquinas. Appeal for ethical guidance was made to God (be that Islamic, Judaic or Christian). Its time span is from around the time of the fall of Old Rome in 410 CE, to around the time of the fall of New Rome—Constantinople, in 1453 CE.

In the *Modern* period, commencing sometime after the Renaissance and the Reformation, rationality, reason, impartiality, empiricism, subjectivity, secularity, humanism and individualism began to replace religious duty. Science and reason replaced religious dogma as authority (hence "the enlightenment" as a descriptive

© The Author(s) 2017
P. Walker and T. Lovat, *Life and Death Decisions in the Clinical Setting*,
SpringerBriefs in Ethics, DOI 10.1007/978-981-10-4301-7_2

term). A morality of 'law, command, duty and obligation' (McCoy 2004), came to replace the external motivations originating in the Classical and Medieval periods. It was in the modern era of rationalism that, in deciding what I should do, there developed three ethical frameworks or groups of guidelines. The origins of each can however be traced to the Classical and Medieval eras (Walker and Lovat 2016). They are the *deontological* framework, the *teleological* framework, and the *virtue ethics* framework. They are considered *substantive*—that is, they are stand-alone sets of guidelines. In brief, deontological ethics focuses on the nature of the Act, and thus its framework is largely independent of situation or context. Teleological ethics focuses on the consequences of the Act, and thus its framework is at least in part conditional upon the situation or context. Virtue ethics focuses on the character of the agent. As well, there persisted a theist framework in the Islamic-Judaeo-Christian tradition. The theist frameworks are aware of an hereafter.

Principles derived from these frameworks, and modifications proposed to them, are foundation stones which have guided ethical decision making, including in clinical situations. Specifically to guide decision making in medical ethics, four principles were distilled from deontology and teleology. These are autonomy, beneficence, non-maleficence, and justice. They are considered in Chap. 3.

To illustrate and explain the ethical frameworks of deontology, teleology, and virtue ethics, we will consider trolley cars (trams) speeding out of control.

2.2   The Ethics of Trolley Cars Speeding Out of Control

Consider a runaway trolley car (tram), which is heading towards five workers on the track ahead, all of whom will be killed if it crashes into them (Foot 1967).

Scenario A: There is a side-track upon which one worker is standing. You are beside the switch which diverts the tram to the side track, which will kill the one worker. Would you flip the switch?

Scenario B: You are on a bridge over the track which the runaway tram must pass under on its way to the five workers. Standing beside you is a large worker, who, if he was pushed off the bridge in front of the tram would himself be killed—but would stop the tram before it reached the five workers. Would you push him off the bridge?

A large number of studies suggest that, regardless of age, ethnicity, religious background, general knowledge and specific moral philosophical knowledge, in Scenario A 89% of respondents would flick the switch. Yet in Scenario B 89% of respondents would not push the large man off the bridge (Hauser et al. 2007).

For Bystander A, the decision to flick the switch so that the out-of-control tram is diverted away from the five workers on the main track, towards the one worker on the side track, arguably, means that Bystander A chooses to save the five workers, at the expense of the one worker. This is the ethical decision making framework known as utilitarianism, consequentialism, or teleology. Bystander B considers the same potential end—the death of the one but preservation of the five—but achieved by different means. When Bystander B chooses not to push the large man onto the tracks, and so sacrifice him in order to save the five workers, this arguably means that for Bystander B, there is something intrinsically or inherently wrong in deliberately pushing the one onto the track below. This is the ethical decision making framework known as intrinsic, categorical, duty-based, or deontological.

2.3 Deontology

The name of Immanuel Kant is associated with the development of the framework of deontological ethics, which as just noted, is also known as categorical, intrinsic, duty-based. Under this framework the moral permissibility of an Action depends upon the intrinsic nature of that Action. Whether an act is morally good or not is dependent upon its concordance with a set of rules or principles, independent of its consequences. For example, torturing a terrorist who knows the location of a large explosive device under a city, is always morally wrong, no matter what the potential consequences of the bomb exploding are.

More formally, two components are important. First, we may make choices to act, only if the rule on which we base our decision to act, can be generalised so that it universally applies to all persons in all circumstances. This is known as Kant's categorical imperative. Second, we should act so that we treat persons always as an end, never merely as a means to an end. This is known as Kant's practical imperative. This means that humans should be considered as intrinsically valuable, with intrinsic dignity. Persons are unable to be compared (that is they are said to be incommensurable) in terms of their physical infirmity, their intellectual incapacity, or their contribution to society. The intrinsic value of a person precludes comparison with any other person, or indeed within oneself at different times or in different states of illness or infirmity.

Deontological frameworks make personal moral integrity paramount. Whatever the consequences, we cannot lie, despite the fact that innocents might thus be saved. In moral conflicts (for example, stealing), Kant would apply both his categorical imperative (stealing another person's property cannot be generalised) as well as his practical imperative (others should be treated as ends in themselves, not as a means to *our* ends). In clinical practice, an example could be that of a transplant physician who has tested an extended family for suitability for bone marrow transplantation for a 3 year old child's incurable leukaemia. He finds one uncle a perfect match. He is the only match. However the uncle is HIV positive after unprotected male sex, about which his wife knows nothing. The family of the child ask the physician whether he has found a match. Under the deontological framework as originally conceived, it is the clinician's duty never to lie—to always speak the truth, regardless of the situation or of the consequences of speaking the truth. Consequences here might well include marital break-up, family fragmentation, job loss and social isolation. A second example is that if some parents choose not to immunise their children, they avoid all potential risks from immunisation of their children, but will still share the 'herd immunity' deriving from the vast majority of parents taking the risks of immunisation. While clearly rational for some to act this way, nonetheless, is it morally permissible? Arguably such parents are using other children (who take the risks) as a means to their own ends (protecting their own child). As a third, more troublesome example, consider the situation when a case of non-paternity in a traditional Muslim family is discovered via genetic incompatibility of the child with the father. While truth-speaking is a deontological maxim, in this situation the consequences of speaking the truth will likely include execution of the mother, abandonment of the daughter, and marriage ineligibility for the mother's sisters (Gray 2015).

Deontological constraints are often framed in the negative, and are narrowly-framed. While you must not lie, you may withhold a truth. The crux is that you, as a morally good agent, must not do an act—not that the act must not happen. Consequences need not be considered in moral decision making under this framework. Consequences may be reasonably foreseeable, but if they are not the means or the end which the agent aims at, then no deontological constraints are breached (Davis 1993). Consider again the deontological rule "do not lie". A murderer comes to the door and asks you whether the victim he is seeking is hiding in your house. The victim is, and under this framework, you must not lie, so you agree that the victim is hiding inside. Since harming the victim is not your chosen end, no deontological rule is breached.

At this point it is important to recognise that there exist in many ethically dilemmatic situations, a hierarchy of duties. It is entirely consistent from a deontologist's perspective that conflicts of *prima facie* duties (that is, obligation/s that, ordinarily, must be fulfilled) exist, and should be resolved by using our best judgement to choose the actual action through which the most-good will be achieved. Stopping to assist at an accident is a higher duty than having coffee (especially if we have a skill-set useful to render medical assistance), despite promising to meet to have coffee. Another example would be the decision to over-ride the moral precept 'thou shalt not kill' in order to shoot a soldier who is burning to an inevitable death in unimaginable pain, but whom we cannot physically approach to give morphine to relieve that pain. It is in fact the duty of non-maleficence to which WD Ross ascribes the primary responsibility for us to follow. It is the only duty expressed in the negative, but in this form, recognition of our duty not to harm others is the basis of the secondary duty of beneficence. From this perspective, non-maleficence is more binding than beneficence. It is not good to steal from one in order to give alms to another, or to kill one in order to keep another alive (Ross 2002). Similarly, familiar to all medical clinicians, *primum non nocere*—first do no harm.

We now turn to a consideration of the question as to how, from a philosophical perspective, people may be allowed to die in Intensive Care Units. Since modern life support technologies are very effective in maintaining our physiological functions, as we have argued elsewhere (Walker and Lovat 2014) there needs to be a robust philosophical basis for decision making in relation to the discontinuation of life-support technology. We approach this via first considering what it is to be a human person.

2.3.1 Personhood

Issues around life, moral agency, and personhood are fundamental moral philosophical predicates to decision making in clinical situations. Life can be understood at two levels. The first is biological or physiological life—characterised by beating of the heart, ventilation of the lungs, digestion, nervous activity, *inter alia*. The second is life in the sense of personhood, and what may be termed the meaningfulness of life. Jeff McMahan posits that we must thus have two corresponding understandings of death (McMahan 2003). The first is the death of the human biological organism. The second is the death, as ceasing-to-exist (here on earth at least), of the human person.

Criteria vary by which to assign personhood to a human being. Some would suggest that 'the mere fact that a being is "human born" provides a strong reason for according it the same status as other humans' (Scanlon 1998; Krishna and Kumar 2013), in which case physiological human life innately confers personhood. Against this is the traditional understanding of personhood that entails moral agency, autonomy, rationality and cognition, linguistic ability, and self-awareness. The traditional emphasis on what we will loosely proscribe as "rationality" is based largely on the Cartesian duality between mind and body—in the sense here, of regarding rationality (especially in the Kantian sense of autonomy) as a separable notion from that denoting a biological human being. Under these criteria, certain members of the species, *Homo sapiens*, can be denied personhood. These include anencephalics (those with absence of a major portion of the brain), infants, young children, the intellectually handicapped, those psychotic, demented and those in a persistent coma (Gert 2004; Engelhardt 1996; Fletcher 1998; O'Donovan 1998; DeGrazia 2006; Warren 1973; Hellston 2000). This approach, however, fails to recognise that even 'our most intellectual thoughts are not independent of either our emotions or our relations to other people' (Mathews 2012). First, we argue that our person is something which develops in the course of our life, predicated upon experiences which, even when not consciously recollected, are incorporated into our habitual bodily responses. We rely on intersubjective bonds which follow from the phenomenological understanding touched upon in Sect. 1.4, of our inter-connectedness, of being in the world of lived experience. We are embedded in a shared world—our experience of which results in our existence in a socio-cultural habitus even when we may not be explicitly aware of it. In other words, our 'humanity which is worthy of moral respect … is located not only in our rational capacities, but in all levels of our being as embodied human subjects' (Mathews 2012), and thus our identities exist within the context of relationships. Second, a mature understanding of self-awareness recognises an empathic relationship with others. '[T]he Cartesian (and Kantian) conception of a person, as a kind of disembodied … thinker and decision-maker' (Mathews 2012) is a limited and partial abstraction from the whole human person. Recognition of vulnerability should be as much a focus of moral concern as our rationality. Rather than a metaphysical derivation, clinical moral decision making requires a reflective phenomenological understanding of human beings as persons, 'which, in Husserl's words, gets "back to the things themselves" as we actually experience them' (Mathews 2012). Martin Heidegger too points to an understanding of our personhood, our essence, as situated-in-the-world with others. More directly, Tom Kitwood argues that personhood is a standing or status bestowed upon human beings, by other human beings, in the context of relationships (Kitwood 1997, 2001). There is also a temporal element to personhood. Paul Ricoeur understands that we are who we are, despite the fact that thoughts, memories, and character traits change over time, because we are 'anchored in sameness by virtue of the temporal connections between past, present, and future' (Mackenzie 2001). This is a more nuanced understanding of the reality of our human condition, allowing the neurologically damaged to be treated in morally good ways by staff (acting as moral agents themselves), caring for them because these staff empathise with their shared humanity as fellow-sufferers.

Thus we need a robust philosophical underpinning for how to approach end-of-life decisions in ICU. It is insufficient to deny personhood solely because the criteria for "rationality" have been lost. Our understanding of autonomy, in a properly authentic sense, should not focus on an egotistical individualist autonomy but should include an awareness of relationships. By doing so, autonomy should be strengthened, not weakened, by the reality of our existence in a world of others. A significant part of the reality for the relatives may well be an emotional clouding as to what the best course of action is. ICU staff, who bring valuable expertise about the clinical condition and its prognosis, may be trained in dialogical methods, and when they recognise the fragility of the critically-ill patient, as well as the wider community of others, are then well-positioned to offer guidance. Thus, we contend, based upon these relational conceptions of personhood and autonomy, that the moral decisions to be made, as the end-of-life approaches, are set in the context of this particular individual patient, considering their lived socio-cultural experience and the relationships they have, as well as the underlying clinical problem and stage of its natural history.

2.3.2 Death in Intensive Care

Traditionally, three constructs may be considered in making moral decisions at the end-of-life. These are the Principle of Double Effect, the Principle of Doing versus Allowing, and the Principle of Ordinary Care versus Extra-Ordinary Care.

The first principle, Double Effect, is derived from Aquinas' *Summa Theologica*. 'Nothing hinders one act from having two effects, only one of which is intended, while the other is beside the intention. Now moral acts take their species according to what is intended, and not according to what is beside the intention' (Aquinas 1947). As a principle for moral decision making, double effect is applicable when a single action can have two (or more) outcomes—one (or more) good and one (or more) harmful. Examples include harming an attacker in self-defence, and wartime dropping of bombs on military targets with foreseeable but unintended civilian casualties. A commonly cited clinical example is minimizing suffering by giving opiate pain relief, which foreseeably depresses spontaneous ventilation and so shortens life (a scenario however for which of course the therapeutic margin in palliative care is actually not so narrow). The primary outcome aimed at is to relieve suffering. The secondary outcome is ventilatory depression. Now recall the trolley-car dilemma above. Some respondents to this problem call upon the principle of double effect. That is, they say that their primary intention is to divert the trolley away from the five; with the secondary unintended (albeit foreseeable) outcome being the death of the one. Consider for a moment that the Principle of Double Effect is active in the minds of both Bystanders A and B. Bystander A flips the switch, the train is diverted towards the one on the track so the five are saved, but then the one hears the train and steps off the track. The five are still saved. Whereas, if Bystander B pushes the large man and he survives the fall and jumps to safety, the train is not stopped by his body, and the five are lost. Even though both Bystanders A and B might both base their decisions on double effect, for Bystander A the one might escape after the five are saved; whereas, for Bystander B, the one must die in order to save the five—this death is, in fact, necessary, and so (arguably) is intended. Those who base their ethical decision making on the consequences which follow the Action (who are termed teleologists, and are discussed in the next section) would generally reject the notion that if two acts have the same actual or foreseeable outcome, they can vary in their moral permissibility. They therefore have difficulty with the principle of double effect.

The second principle has become known as Doing versus Allowing (Quinn 1989). As a basis for ethical decision making, it distinguishes between acting and not-acting. This distinction applies to, for example, killing someone by actively holding their head under water, versus failing to rescue a person who is drowning. A clinical example is active euthanasia by lethal injection, versus passive euthanasia by switching off life-support technology or by withholding antibiotics for pneumonia. In the context of euthanasia, James Rachels argues that once the decision to not prolong life and to avoid suffering has been made, there is no ethically significant difference between active euthanasia and passive euthanasia. Put another way, he sees no practical difference, ethically, between killing and letting die. Rather, the moral difference hinges on motive and intent.

Traditionally, the third principle is 'Ordinary' means versus 'Extra-ordinary' means. This can be summarised as: ordinary means are proportionate means, while extraordinary means are disproportionate means. This principle looks to the benefits which might accrue to the individual patient for whom it is being considered, compared with the difficulties and risks. This approach has been refined as 'Benefit versus Burden analysis'. It considers the potential benefits of a proposed treatment plan, and compares those with the potential burdens of the proposed treatment plan.

It adopts a teleological approach in that it determines utility and disutility, and then derives nett utility. To be useful in guiding decisions as to whether treatments should be commenced or withheld, it should be contextualised to the individual patient's situation. Instituting potentially life-maintaining interventions in, for example, severely syndromal neonates (including mechanical ventilation or Extra Corporeal Membrane Oxygenation), may well place an unreasonable burden of discomfort on the neonate, and a burden of prolonged suffering on the family, not outweighed by the benefits likely to flow from that intervention in terms of the quality of life which the child might achieve. Inevitably, it involves both the clinician and the patient making their best estimation of the likely benefits and burdens attached to the proposed treatment. The balancing of benefits versus burdens can only be meaningfully approached by having a dialogue amongst those on whom the decision impacts—namely, the patient and their family, and also the clinicians who will be advising what the treatment options, and what the risks and benefits, are likely to be.

In approaching end-of-life decision making, we favour benefit versus burden analysis. This is because, not only is it necessary to anchor the end-of-life decision making in the unique context of the individual patient, but also because this framework looks to the Good which may accrue to the patient and their family in their actual situation. The aim is to maximise the patient's good as a priority, allowing for the actual real clinical situation they are in, and not in isolation from others who are in relationship with this patient.

2.4 Teleology

The names of Jeremy Bentham and John Stuart Mill are associated with the development of the framework of teleological ethics. Under this framework the moral permissibility of an Action is determined by its outcome. Consequentialists are concerned with making moral decisions based upon the outcome or potential outcome which follows the Act. Whether an Act is morally good or not depends upon whether it brings about the best consequences, independent of the reasons for acting. For example, torturing a terrorist may be morally right, depending upon the result. In a medical example, consider the situation where there is only one dialysis machine available in a hospital. Six patients need renal dialysis to stay alive—five patients need it daily but briefly but the sixth requires it for a whole day, precluding the use of the machine by the five. Under a teleological framework, the five are chosen for dialysis over the one—who does not receive dialysis.

Utilitarians focus upon the greatest good for the greatest number, so do not specifically consider distinctions among individuals. This framework requires summation of the degrees of tendency to goodness for each individual, in regard to whom there is goodness for the whole, then offsetting this by the summation of the tendency to badness for the whole, and then taking the resultant balance as the tendency to goodness of the act for the community as a whole. Focusing upon the nett balance of utility relieves them of the need to concern themselves with individual allocations of good (or bad) things. Specifically, the greater gains of some can compensate for the lesser gains of others.

Returning to the runaway tram once more, there has now been a tram accident.

In Accident 1, the result is that 4 patients are moderately injured but will likely live with sufficient resuscitation; while 1 is severely injured and will likely take most of the resuscitative team's resources, with an uncertain outcome even then.

In Accident 2, 4 workers are severely injured needing transplants to survive, and 1 healthy patient is in the Outpatients Department next door, for a routine check-up.

Following Accident 1, the patients are triaged into categories of urgency for medical intervention. From its origin in military mass casualty situations, the principles of triage have been adopted for the medical management of disasters and emergency medicine. Although Kant (and others of course) have argued that humans should be seen, always as ends in themselves, and intrinsically valuable regardless of functional ability, principles of triage are inherently utilitarian, directing limited resources to salvage the greatest number of casualties. Triage in the civilian setting aims to separate those with a potential to be saved by immediate treatment, and prioritises these over others who do not appear as likely to be able to be saved whatever treatment is offered, and over those with lesser injuries. However, an exception is usually made for injured emergency staff who will be able to return to saving others—and so they will generally also be prioritised over non-medical personnel. Similarly, they may be given priority immunisations in a pan endemic in order to allow them to stay at work and continue to immunise others. Triage of the wounded in military situations however allows for, indeed encourages (Beebe and DeBakey 1952), prioritisation so that less-wounded soldiers may be treated first in order to get them back into their defensive positions, so as to prevent the perimeter being over-run. Military capability is seen as a greater good than individual patient care. Using these principles, in North Africa during World War II a decision was made to allocate scarce penicillin injections to those infected with gonorrhoea, rather than to those infected after war injuries, because those treated for gonorrhoea would return to battle much more quickly (Beecher 1969). Following Accident 1, it is likely that resuscitation will be offered to the four moderately injured patients, at the expense of the one severely injured patient. However, extending the principles of military triage to prioritise allocation of intensive care beds to those most quickly able to return to contributing to society, is not likely to be considered appropriate by most clinicians.

Following Accident 2, although the teleological framework allows harvesting 4 organs from the well patient in for a check-up, in order to benefit the 4 severely injured patients who need them, in clinical practice this is also unlikely to be acceptable. That is, there are clearly some things which, intrinsically, should not be done, even for the greater good. That is, the "greatest good for the greatest number" and the "ends-justify-the-means" ethics need to be evaluated with care. Minority groups may well be disadvantaged as the nett utility calculation benefits the numerically larger group. In another clinical example, consider the allocation of limited health care resources in the provision of acute care (for example, to treat children with poliomyelitis) versus preventative care (immunisation of infants against poliomyelitis). The greater long term nett gain will likely accrue from immunisation, but, according to the strictly adhered to principles of teleological ethics, those individuals needing acute care, must be refused that acute treatment in order to achieve the greatest good for the greatest number.

Consider also, a former Olympic skier who, because of financial hardships, has been unable to ski for several years. She is seven months pregnant when she and her husband win an all-expenses-paid holiday to go skiing for three months. She wants an abortion so she can have that ski holiday. These two outcomes cannot easily be compared. The outcome for the mother, following from the abortion, is that she can have a holiday. The outcome for the foetus is death. Thus, a hierarchical ranking of consequences (including present consequences versus future consequences) is just as important as a hierarchy of duties.

2.5 Virtue Ethics

Virtue ethics is concerned with making moral decisions in the Aristotelian sense of according with virtue, rather than according to rules or consequences. Under a virtue ethics framework, the development of personal moral virtue allows for morally Good decisions to be made. It could be said that this is an emphasis on *being* morally Good rather than on the specifics of *doing* Good.

Set against the adversarial background of debate between teleological ethics and deontological ethics throughout the eighteenth, nineteenth, and early twentieth centuries, virtue ethics has had resurgence in the latter half of the twentieth century. As currently iterated, most virtue ethical frameworks are neo-Aristotelian in spirit. Aristotle spoke to the virtuous mean between excessiveness and insufficiency. For example, consider courage. Too little is cowardice. Too much is recklessness. Consider 'pleasantness'. A deficiency in a person's state of character with regard to pleasantness leads to quarrelsome, surly, or unpleasant behaviour. An excess is obsequiousness, or flattery. A mean state of character is friendly or pleasant. Similarly, dignity lies between servility and selfishness; and so on. Virtue is the mean between these two extremes.

A virtuous person seeks preferentially after intrinsic goodness (beneficence, generosity, honesty, courage) rather than instrumental goodness (fame, money, power). This is combined with sensitivity as to when and where a moral issue exists and an inherent motivation to act in a virtuous manner. While there can be erudite discussion about what is virtue, what are the virtues, and how we might aspire to be virtuous, we suggest that the essence is that 'good character guides right action: the ethical aim is to form oneself as a good person, and a well-formed person both knows how to act rightly and will habitually choose to do so' (Balousek 2014). Under a virtue ethics framework, moral decisions are made by well-informed, habitually good people who consider the individual situation, allow for their earlier experiences and knowledge, and make the morally best decision they can in that situation, learning from it for future situations, for intrinsically good reasons.

Darrin Balousek argues that a virtue ethics framework, in the context of guiding decisions about whether to allow performance-enhancing drugs in professional sport, would have us ask two questions (Balousek 2014). The first is whether a decision in favour or against would foster habits which are formative of good character in the athletes, the team owners, and the fans of the game, and which thus enhance *eudaimonia*—the Greek notion of *flourishing*. The second is whether the decision for or against promotes excellence in that sporting activity (which Balousek ascribes to excellence in performance, rather than to winning per se), and which thus promotes the intrinsic good of that activity.

Edmund Pellegrino restates the Aristotelian definition of Virtue in the setting of the physician *qua* (as being) physician, striving to maximise the Good of the patient via a right and good healing action or decision. He argues that the good of the patient has been the foundation of morality in clinical medicine since antiquity (Pellegrino 1985). It is the ultimate arbiter in clinical decisions from a moral perspective. He proposes a hierarchy of four Goods of the patient. The highest Good is the *summum bonum*. The least Good in the hierarchy is the biomedical or techno-medical Good. This is an instrumental good which follows from the correct diagnosis, the correct drug in the correct dose or the correct operation, all in a technical sense. Next in the hierarchy is the perceptual Good of the patient, how she understands the clinical situation and treatment options, and how she wants to proceed. While competent, only the patient can judge what is most perceptually good amongst treatment options. For example, one of the risks of radical prostatectomy for prostatic cancer is incontinence and erectile dysfunction. Although the statistical risks of progression of the cancer can be imparted, 'the precise meaning of the cancer in [his] life will be unique to him' (Cowley 2012). Next is the Good of the patient as a human person, which Pellegrino founds upon autonomy. If not competent to choose, then another acts as a surrogate.

Each participant in the decision-making process in a clinical dilemma is trying to make a decision for the Good of the patient, while allowing that the participants will have different understandings of the patient's Good. This is similar to seeking the patient's "best interests". "Best interests" may be in common usage by clinicians, but this book argues that this term is both vague, and somewhat simplistic. It should be more rigorously restated in terms of the *Good* of the patient, allowing recourse to the four Goods proposed by Pellegrino.

In postulating a model for 'the right moral attitude' in clinicians, Petra Gelhaus has looked at (the capacity for) empathy (Gelhaus 2011a), compassion (the adequate professional inner attitude) (Gelhaus 2011b), and care (the active *praxis* of clinical medicine) (Gelhaus 2012). Thus she proposes what she terms the normative image of a good physician. This book argues that both empathy and compassion require three things from the clinician. These are: first, an awareness of others; second, an awareness of their suffering when they become patients; and third, recognition of the inherent vulnerability of patients *qua* patients. This awareness may be located in the metaphysical space, and underlines once again the insights into moral decision making in clinical situations, which phenomenology offers. Phenomenology places the patient in their actual reality, as the starting point for their care.

This approach is also of relevance to clinical decision making as it applies to the care of children. Virtuous parents have a moral authority to make decisions for their children. This may be either as proxy decision-makers for their child ("which school would my child like to attend?"), or as autonomous decision-makers about parenting itself ("which school will the parent decide to send their child, considering cost, belief system, etc."). In this understanding, parents have a right to raise their children in accord with their own values. For example, parents are able to over-ride their child's best interests by enforcing their parental value of "share your toys with your siblings", and in so-doing, seek the best interest of the whole family. In clinical situations where there is disagreement between clinicians and parents when, for example, parents choose not to allow blood transfusions or chemotherapy for their child, a more practical tool is required. Lynn Gillam draws upon the harm principle and proposes that while parents have an obligation to maximise the well-being of their children, they have an absolute obligation not to cause significant harm. She uses the term "Zone of Parental Discretion" to describe an ethically-protected range for decisions which, while they may not be the very best for their child, do not cause significant harm (Gillam 2014). It may be possible to grade parental decisions in clinical situations as optimal, sub-optimal but reasonable, and harmful. Decisions in this last group may be over-ruled. Parents are able to exercise their parental autonomy only up to a point, the point of significant harm.

Finally, although attractive, especially in medicine, we suggest that in actual clinical situations, no virtue ethicist could be expected to know what constitutes the multiple goods to be enhanced, without a dialogue amongst those affected.

2.6 Islamic-Judaeo-Christian Influences

The question we pose here is whether the monotheistic Faiths of the Islamic, Judaic and Christian God constitute a substantive moral framework comparable to the secular normative ethical frameworks of deontology, teleology, and virtue ethics; or whether their importance is recognised solely in terms of their influence on the secular normative ethical frameworks. While historically a substantive moral code, which reached its zenith in the medieval era, arguably the Islamic-Judaeo-Christian tradition is no longer a moral framework with normative force. Although no-one would be likely to suggest that these monotheistic Faiths have a monopoly on Good and Goodness, there has been a substantial body of work published which addresses the questions of why and how to make moral decisions from this perspective. Warnock posits that 'though … moral philosophy … has been secularised, it is almost impossible to think about the origins and development of morality itself without thinking about its interconnections with religion' (Warnock 2004).

Each of the traditions of Judaism, Islam and Christianity begin with revelations originating from a divine source and recorded as sacred texts in the first five books of the Bible—the *Pentateuch*, and in the *Qur'an* and the New Testament. To the Hebraic *Tanakh*, is added the *Qur'an* by Muslims, and the New Testament by Christians. Equality before God is a tenet of all three traditions. Each tradition then allows, indeed fosters, interpretation of the divine scriptures in the practical moral decision making of humans by the respective community of the faithful. Each tradition is characterised by an ethical impact and, in turn, practical action, as the keystone of their faith, be it the practice of the Ten Commandments, Jesus' commandment to love God and neighbour as oneself, or the Five Pillars of Islam. For example, the Bible contributes to foci within Christian Ethics (or Moral Theology) concerned with two basic issues—'how to act from the right motive and how to find what is the right action in particular circumstances' (Preston 1993). Philip Hallie recognised two kinds of ethical rules spread through the Bible—negative rules and positive rules. The chief negative rules were those Moses brought down from Mt Sinai—'Thou shalt not make for yourself an idol ... You shall not murder, Neither shall you commit adultery, Neither shall you steal, Neither shall you bear false witness ..., Neither shall you covet your neighbour's wife ... house, or field' (Deuteronomy 5:8–21). The positive rules include 'learn to do good; seek justice, rescue the oppressed, defend the orphan, plead for the widow' (Isaiah 1:17) and in the Gnostic Gospel of Truth 'Steady the feet of those who stumble and extend your hands to the sick. Feed the hungry and give rest to the weary' (Gospel of Truth 33:1–2) (Valentinus 2007). The negative ethic forbids certain actions; the positive ethic requires certain actions. 'To follow the negative ethic is to be decent, to have clean hands. But to follow the positive ethic, to be one's brother's keeper, is to be more than decent—it is to be active, even aggressive. If the negative ethic is one of decency, the positive one is the ethic of riskful, strenuous nobility' (Hallie 1981).

The Golden Rule of Jesus is 'do unto others as you would have them do unto you' (Mathew 7:12, Luke 6:31). It has a similar existence in most moral traditions. It is aligned with universalizability,and hence in concordance with moral justice; furthermore, it implies reciprocity. One might possess a relatively unsophisticated understanding of ethical principles, yet intuitively apply these understandings successfully to oneself, at least as a sufficient starting point for them to become the basis of an ethical code which one then can apply impartially to others. It is an ethical code which explicitly requires empathically walking in the shoes of the other. It is included in *The Fellowship Pledge* of the American College of Surgeons in the words 'I promise to deal with each patient as I would wish to be dealt with if I were in the patient's position' (American College of Surgeons 2008).

Amongst the three cardinal virtues (1 Corinthians 13:13) *compassion* is most intimate to the Christian Ethic. In John's Gospel the concentration is upon a sophisticated *agape* or 'love' in the sense of unconditional love from God to Man, and amongst men, love of neighbour *as* oneself. As a motivation, *agape* does not give detailed content to ethical decision making (Preston 1993). Rulings were not given by Jesus in a wide variety of different situations which were ethically dilemmatic. The ultimate test for concordance with a Christian Ethic is whether it accords with love of God and love of neighbour. What then is the practical mechanism by which a modern-day Christian might seek to make morally correct decisions? Dietrich Bonhoeffer has an understanding of the essence of the Christian Gospel message and the essential charter of Christianity as 'a deeply personal commitment to strive for the good of others' and to conform 'one's life to serving the betterment of the human race' (Lovat 2012).

In certain clinical situations, for the parents who follow in the Islamic-Judeo-Christian tradition, this framework is very apposite. Consider end-of-life decisions centred on a child with a progressive neurological condition which will ultimately result in death. This child faces the alternative of being placed on a ventilator for several decades and then dying, or being allowed to die immediately. The choice for the parents could be seen as one between being in a better place (heaven) immediately, or prolonged lingering here on earth.

2.7 Summary

We have argued in this chapter that the frameworks we have traditionally appealed to as ethical guides—deontology, teleology, and virtue ethics, have significant shortcomings. Given the multicultural and multi-faith characteristic of our times, these approaches have limitations as substantive frameworks, especially in clinical situations. We propose that moral decision making in clinical contexts should look beyond the established frameworks towards a different approach, identified in the next chapter as Proportionism. The proportionist approach seeks the highest good based upon a virtuous mean or balance-point between a priori rules and an empirical "greatest good for the greatest number" utilitarian approach. Practical application to moral decision making in clinical situations is based upon a process of inclusive, non-coercive and self-reflective dialogue within the community affected. This in turn is founded upon Habermas' communicative action incorporating his discourse theory of morality, aimed at reaching an unforced consensus, predicated upon language aimed at establishing an ideal speech situation.

References

American College of Surgeons. 2008. Fellowship pledge. http://www.facs.org/fellows_info/statements/stonprin.html#fp. Accessed April 28, 2014.

Aquinas, T. 1947. Summa theologica. *Treatise on the cardinal virtues*, Q64 Art67. Benziger Bros.

Balousek, D. 2014. Professional baseball and performance-enhancing drugs. *Philosophy Now* 102 (May–June): 14.

Beebe, G.W., and M.E. DeBakey. 1952. *Battle casualties: Incidence, mortality, and logistic considerations*. Springfield Illinois: Charles C Thomas.

Beecher, H.K. 1969. Scarce resources and medical advancement. *Daedalus* 98 (2): 275–313.

Cowley, C. (ed.). 2012. *Reconceiving medical ethics. Continuum studies in philosophy.* London: Continuum International Publishing.

Davis, N.A. 1993. Contemporary deontology. In *A companion to ethics. Blackwell companions to philosophy,* vol. 2, ed. P. Singer, 205–218. Malden, MA: Blackwell.

DeGrazia, D. 2006. On the question of personhood beyond *Homo sapiens.* In *In defence of animals: The second wave,* ed. P. Singer, 40–53. Malden, MA: Wiley-Blackwell Publishing.

Engelhardt, H.T. 1996. *The foundations of bioethics,* 2nd ed. New York: Oxford University Press.

Fletcher, J. 1998. Four indicators of humanhood—The enquiry matures. In *On moral medicine: Theological perspectives in medical ethics,* 2nd ed, ed. S.E. Lammers, and A. Verhey, 376–379. Grand Rapids, MI: William B Eerdmans Publishing.

Foot, P. 1967. The problem of abortion and the doctrine of double effect. *Oxford Review* 5: 1–7.

Gelhaus, P. 2011a. The desired moral attitude of the physician: (I) Empathy. *Medicine, Health Care and Philosophy* 15 (2): 1–11. doi:10.1007/s11019-011-9366-4.

Gelhaus, P. 2011b. The desired moral attitude of the physician: (II) Compassion. *Medicine, Health Care and Philosophy* 15 (4): 1–14. doi:10.1007/s11019-011-9368-2.

Gelhaus, P. 2012. The desired moral attitude of the physician: (III) Care. *Medicine, Health Care and Philosophy* 16 (2): 125–139. doi:10.1007/s11019-012-9380-1.

Gert, B. 2004. *Common morality: Deciding what to do.* Oxford: Oxford University Press.

Gillam, L. 2014. When parents and doctors disagree about medical treatment for a child: The ethics of decision making. In The New Zealand Bioethics Conference, Dunedin, Otago, 24 January 2014.

Gray, B. 2015. Culturally competent bioethics: Analysis of a case study. *Journal of Bioethical Inquiry* 12 (2): 361–367. doi:10.1007/s11673-015-9636-6.

Hallie, P. 1981. From cruelty to goodness. *The Hastings Center Report* 11 (3): 23–28.

Hauser, M., Cushman, F., Young, L., Kang-Xing, J., and Mikhail, J. 2007. A dissociation between moral judgements and justifications. *Mind and Language* 22 (1): 1–21.

Hellston, S.K. 2000. Towards an alternative approach to personhood in the end of life questions. *Theoretical Medical Bioethics* 21 (6): 515–536.

Kitwood, T. 1997, 2001. On being a person. In *Dementia reconsidered,* 8. Buckingham, UK: Open University Press.

Krishna, R., and L. Kumar. 2013. Accounting for personhood in palliative sedation: The ring theory of personhood. *Medical Humanities* 40 (1): 17–21. doi:10.1136/medhum-2013-010368.

Lovat, T. 2012. Bonhoeffer: Interfaith theologian and practical mystic. *Pacifica* 25: 177–189.

Mackenzie, C. 2001. On bodily autonomy. In *Handbook of phenomenology and medicine. Philosophy and medicine,* vol. 68, ed. S.K. Toombs, 417–440. Dordrecht: Kluwer Academic Publishers.

Mathews, E. 2012. Old age and dependency. In *Reconceiving medical ethics. Bloomsbury studies in philosophy,* 1st ed, ed. C. Cowley, 59–71. London: Bloomsbury Academic.

McCoy, A. 2004. *An intelligent person's guide to Christian ethics.* London: Continuum.

McMahan, J. 2003. *The ethics of killing: Problems at the margins of life.* Oxford: Oxford University Press.

O'Donovan, O. 1998. Again: Who is a person. In *On moral medicine: Theological perspectives in medical ethics,* 2nd ed, ed. S.E. Lammers, and A. Verhey, 380–386. Grand Rapids, MI: William B Eerdmans Publishing.

Pellegrino, E. 1985. Moral choice, the good of the patient, and the patient's good. In *Ethics and critical care medicine. Philosophy and medicine,* vol. 19, ed. J.C. Moskop, and L. Kopelman, 117–138. Dordrecht: D Reidel Publishing.

Post, S.G. 2004. *Encyclopedia of bioethics,* vol. 1, 3rd ed. Macmillan Reference USA.

Preston, R. 1993. Christian ethics. In *A companion to ethics. Blackwell companions to philosophy,* vol. 2, ed. P. Singer, 91–105. Malden, MA: Blackwell.

Quinn, W.S. 1989. Actions, intentions, and consequences: The doctrine of doing and allowing. *Philos Review* XCVIII (3): 287–312.

Ross, D. 2002. *The right and the good. British moral philosophers,* 2nd ed. Oxford: Clarendon Press.

Scanlon, T.M. 1998. *What we owe to each other*. Cambridge, MA: Belknap Press of Harvard University Press.

Valentinus. 2007. The Gospel of truth. In *The secret Gospels of Jesus*, ed. M. Meyer, 89–112. London: Darton, Longman and Todd Ltd.

Walker, P., and T. Lovat. 2014. Concepts of personhood and autonomy as they apply to end-of-life decisions in intensive care. *Medicine, Health Care and Philosophy* 18 (3): 309–315. doi:10.1007/s11019-014-9604-7.

Walker, P., and T. Lovat. 2016. Towards a proportionate approach to moral decision making in medicine. *Ethics and Medicine* 32 (3): 153–161.

Warnock, M. 2004. *An intelligent person's guide to ethics*. London: Overlook Books.

Warren, M.A. 1973. On the moral and legal status of abortion. *The Monist* 57 (1): 43–61. doi:10.5840/monist197357133.

Chapter 3
Balancing Old and New Approaches: Principlism Versus Proportionism

3.1 Principlism

Appeal to substantive normative moral frameworks to guide moral decision making has been complicated by uncertainty about which normative framework to apply; and if more than one may be applicable, with what weighting each should each be applied. Thus medical ethicists made attempts to simplify the decision making process in clinical situations. The idea was to determine a smaller number of principles, and rules derived from them, which would usefully guide moral decision making by clinicians. Tom Beauchamp and James Childress, in their *Principles of Biomedical Ethics*, propose a set of four principles as an analytical framework to express the norms of common morality as they see them applying to medical ethics (Beauchamp and Childress 2013). The four principles are *respect for autonomy*, *non-maleficence*, *beneficence*, and *justice*. Very influential in medical ethics they originate from the normative ethical frameworks discussed in Chap. 2. Although they may be seen as a group of intermediate or mid-level practical guidelines 'located just below theories and just above rules' (Clouser and Gert 1990); at a minimum, they are popular as a common moral language amongst clinicians, and so are in quite widespread use.

3.1.1 Autonomy

Although the four principles are conceived as having equal weighting, de facto, the principle of respect for autonomy is considered the dominant principle (Raanan 2003; Wolpe 1998). This originated largely as a reaction against clinical and research "experiments" during the 20th century. These were those of the Nazis in World War II (Weindling 2001), and the Tuskegee syphilis experiment (Jones James and the Tuskegee Institute 1981, 1993), amongst others. It also followed

© The Author(s) 2017
P. Walker and T. Lovat, *Life and Death Decisions in the Clinical Setting*,
SpringerBriefs in Ethics, DOI 10.1007/978-981-10-4301-7_3

increasing awareness of the rights of individuals in wider society. Three traditional concepts of *autonomy*, as they might apply to moral decision making in clinical situations, may be considered—derived from the understandings of Immanuel Kant, John Stuart Mill, and Edmund Pellegrino. These traditional concepts of autonomy however, are problematic in practical application. A more nuanced understanding may be more applicable to clinical moral decision making.

The moral philosophical sense of *autonomy* has its grounding in the writings of Kant. It leans towards his categorical imperative of conforming one's will to the self-legislated rational dictates of the moral law (Larmore 2008; Deligiorgi 2012). Personal autonomy leans towards the sense inherent within Kant's practical imperative—that all persons have intrinsic moral worth, and that it is impermissible to treat another human as a mere means to an end. As applied to patients, it thus encompasses the concept that the individual makes decisions, free from the controlling influence of others, in the vision of their own values, and with adequate understanding of the decision, its necessity and consequences. Mill argued that society should allow individuals the liberty to express their freedom, even to harm themselves, provided they do not harm others. Specifically, to forcibly intervene for 'his own good, either physical or moral, is not a sufficient warrant … [nor] because it will be better for him … because it will make him happier, because in the opinion of others to do so would be wise, or even right' and 'over himself, over his own body and mind, the individual is sovereign' (Mill 1952). Pellegrino, in basing his notion of the Good of the patient as a human person, upon patient autonomy, writes that '[w]e cannot override those choices even if they run counter to what we think is good for the patient … even to do what we think is good is to violate his good as a human being' (Pellegrino 2013). Practical concepts in clinical medicine which follow from the principle of autonomy include, amongst others, informed consent, medical confidentiality, and promise-keeping.

Respect for autonomy generally takes priority over beneficence. This means that it is not permissible to coerce or deceive patients, even in their own interests. Medical confidentiality encompasses decisions about whether it is morally permissible to breach patient confidentiality about, for example, a serious infectious disease such as Human Immunodeficiency Virus, in order to protect another who is innocent; or whether to advise genetically-related family-members about a serious genetic diagnosis in the patient; and whether it is permissible to lie in order to engender a better psychological attitude, to avoid depression, and so achieve a better outcome.

Recent biomedical advances render more problematic questions around respect for autonomy in several areas. Consider a patient who has a fully implanted cardiac permanent pacemaker (PPM) (Walker et al. 2014). The PPM senses whether there is a cardiac rhythm and, if not, it is programmed to generate an electrical stimulus. Compare this with a patient with an implanted mechanical cardiac valve. Despite being implanted, it may be argued that the patient with the cardiac valve retains autonomy about whether to continue with this life-maintaining treatment. The patient can effectively withdraw consent for this treatment by discontinuing the prescribed warfarin dose; aware that without warfarin, the valve will likely thrombose and fail. The patient with a PPM has no autonomy to withdraw consent for this life-sustaining treatment. Even if 'tired of life', and seeking death, it is not possible for the patient to switch off or re-programme the PPM herself. The patient cannot withdraw consent for this treatment. In fact, the issue of re-programming will likely only be addressed as permissible or impermissible, if the patient has some other life-threatening illness.

A further example, wherein traditional understandings of autonomy and informed consent are problematic, is seen in the human genome project. In sequencing the child's genome from blood drawn from the mother, the mother's genome must also be sequenced, so that it is clear which is that of the mother and which is that of the child. Thus non-invasive prenatal screening has become a reality. Mother, and child, may find out that they have the genetic code for a serious, but currently asymptomatic disease, for which there is no available intervention. There will also be insurance issues. Does "reproductive autonomy" over-ride the child's and the mother's right *not* to know? (Newson 2013). Similarly, tissue-banking may allow familial diseases to be diagnosed well after the donor has no use for the information. The donor's descendants gave no consent for the information to be looked-for. Future information which may be available is unknown, and arguably, unknowable.

Elsewhere we have argued that we need to move towards a more nuanced understanding of autonomy (Walker and Lovat 2014). In order to further consider autonomy in practical clinical application, consider the once again the critically-ill patient in ICU, potentially with their end-of-life in view. This patient, or their relatives, may claim the right to unlimited technological support, without consideration of feasibility, or costs and limitations to resource availability, on the basis of autonomy. We suggest, however, that this simplified concept of autonomy is insufficient. Because it is derived from Kant's rational self-legislation or Mill's primacy of the rights of the individual, even to the exclusion of staff experienced in critically-ill situations (Pellegrino), it is ultimately egoistic (Deligiorgi 2012), individualistic, and does not require positioning the individual in the world of others (termed *alterity* in philosophy), the world as it actually is. That is, echoing the notion of personhood to which we subscribed in Sect. 2.3.1 rather than existing in isolation, we are in fact 'interdependent, interconnected, and intermingling' with others (Herring and Chau 2007). As well, within the technologically-enhanced milieu of the ICU, humans are fragile, and life-sustaining technologies can be intrusive as well as invasive. An authentic understanding of autonomy recognises the other, those in relationship with the patient, and ethical decisions should therefore be predicated upon this. In our view, a richer account of autonomy in this context implies that the patient connects with their actual situation—namely, that they are in an ICU and gravely ill. In trying to make morally-good decisions in this situation, the quality of the person's future life should become part of the context of the patient—that is, 'seeing patients' ethical dilemmas as grounded in concrete existential situations' (Carel 2011). Put another way, the lived body 'is not a thing, it is a situation' (de Beauvoir 2011). Philosophically, the aim is, first of all, to clarify where the Good lies for this patient, and, second, to seek to maximise that

Good by grounding a full and frank discussion about the quality of life the patient is likely to have, rather than merely its quantity, upon this understanding.

3.1.2 Non-maleficence, Beneficence, Justice

Non-maleficence imposes a negative duty or obligation. That is, a duty or obligation not to inflict harm on a patient, or to minimise the harm inflicted ("harm minimisation"). For example, it is not morally acceptable to sacrifice one person in order to harvest sufficient organs to save five others, despite the obvious utilitarian attraction. A needle however is a necessary pre-requisite to the avoidance of measles. In decision making in clinical situations, clinicians seek the best possible outcome with the least amount of harm. As already noted in Sect. 2.3.2 this is more difficult when there is a level of unpredictability or uncertainty about both the potential outcome and the potential harm. Many interventions, including those which are potentially life-saving, have iatrogenic complications attached. Practical concerns which follow from this principle include the provision of sufficient training to perform the planned procedure in this individual patient, the consideration of side-effects of medication in this individual patient, and the consideration of possible negative effects of participation in research trials. In medicine, this duty is usually regarded as taking precedence over the duty of beneficence.

Beneficence requires that we endeavour to do the best we can for our patients, that is, we try to help them. In the words of Petra Gelhaus, as outlined above in Sect. 2.5 this means empathic, compassionate caring. It also means being aware that our patient is suffering and, as also discussed in Sect. 2.5 aiming in our decision making, to maximise the four goods of the patient which he recognises. We think it reasonable that beneficence is viewable as a continuum. Beneficence implies more than doing the minimum duty required. Writing a referral letter for a patient to a specialist saying "please do the needful" satisfies the regulations but a truly beneficent family doctor will likely record the history, the clinical concern, what investigations have been done, and what response to treatment has been. Doing much more than duty requires is termed "supererogatory", and potentially elevates someone to the level of moral saint or moral hero. For example, rushing into a burning building to rescue someone, or providing urgent medical care by crawling under an overturned train liable to collapse and injure the rescuers would be regarded as supererogatory. Supererogatory actions are not required, and nor is major self-sacrificial altruism, but some clinicians are tempted to overdo their commitment to patients, at the expense of the well-being of themselves, that of their family or even sometimes the patient. Beneficence takes priority over autonomy in emergency situations when the patient's wishes cannot be determined. However, in elective situations, when a clinician acts against the wishes of a rational patient, even for a beneficent reason, it is labelled as paternalism. While it is typically thought to be of a lesser order duty than non-maleficence, beneficence may in fact be more difficult because it requires an active positive intervention, with a necessary but often difficult and time-consuming balancing of risks and benefits. This is especially true when the beneficent clinical intervention is prophylactic, and not without risks, aiming to prevent a disease which has not yet developed. Non-maleficence includes the requirement that clinicians keep up to date with their knowledge and skills, and are motivated in their decision making by the desire to help patients, rather than by any other reason such as financial reward.

Justice includes obligations related to fairness—in terms of the distribution of limited health resources, respect for patient's rights and respect for the Law. Practical concepts which follow from this principle include equal access for all, to limited health care resources. Potential conflicts can arise between patient autonomy to choose a treatment which is likely to be futile, and limited health care resources which are likely to be more beneficial if directed elsewhere. It is necessary to have a conversation about the meaning of the good of the patient in his or her particular context, what values we agree, as a society, are to be ascribed to this concept and how to resolve disagreements about it. Aristotle points out though that to be treated justly does not mean that all must be treated equally—justice involves treating equal persons equally, but treating unequal persons unequally (Aristotle 2000 [III:9 1280a 7–22]). In this sense, treating all people the same is effectively the neutral approach. In some decision-making situations, our patients are not equal, and so might in fact need positive discrimination, the basis for affirmative action. So, in terms of class-placement at school, rather than allocate the front row of seats randomly, alphabetically, or by the drawing of lots, those children with a hearing loss should be chosen to be seated in the front row; this is active discrimination. Also falling under this principle are questions about taking responsibility for the consequences of poor personal health choices, for example, continuing to smoke, or to abuse drugs, where there may be an argument about forfeiture of the right to health care.

3.1.3 The Four Principles in Practice

Critics argue that the four principles (and other similar lists collectively grouped together as the somewhat pejorative *principlism*) act as prompts for 'values worth remembering, but lack deep moral substance and capacity to guide action' (Beauchamp and Childress 2013). We see their value as somewhere between being no more than a mere checklist, of use primarily to those with limited ethical knowledge (Harris 2003); and, viewing them as 'the four moral nucleotides that constitute moral DNA—capable, alone or in combination, of explaining and justifying all the substantive and universalizable moral norms of health care ethics' (Raanan 2003). Lack of a systematic relationship amongst the principles and lack of an explicit priority ranking are seen as particular deficiencies.

Autonomy and beneficence are principles in tension with each other. It is not permissible to coerce or deceive patients, even in their own interests. The principle of beneficence is likely to make appeal to consequences and so be justified teleologically, while the rules of permission are binding independent of consequences, and so are justified deontologically. In a richer account of autonomy, the patient should not simply act upon their own wants or desires. This is because the patient recognises those others with whom they are in relationship, and that the patient connects with their actual situation, which is not only shared with others but which means that moral decision making must be predicated upon the actual reality of their illness situation. Hence, we argue that if the final purpose of medicine is helping patients—via seeking to maximize the Good of the patient—then beneficence must, if not replace, at least come to co-equal autonomy as the *prima facie* first principle.

A number of commentators view the four principles as foundational to a modern medical ethic. At the very least, they have provided a language with which to speak about moral dilemmas. Note however that the well-formed conscience of the moral agent is conspicuously absent from the four principles approach.

3.2 The Proportionist Approach

Aware of the ontological imperative of context, the *proportionist* approach, an approach of proportionate balance, seeks the highest good based upon seeking the virtuous mean or balance-point between a priori rules and empirical "greatest good for the greatest number" utilitarian calculations, with, as its starting point, the actual reality of the patient and their situation (Walker and Lovat 2016). This approach is put into practice via a dialogue amongst the clinician, the patient and their family, which seeks to achieve a consensual decision about wherein lies the greatest good for the patient. We term this approach proportionist because the ensuing dialogue seeks to understand the values which are important to the patient, their family and others in the dialogue, in order to give them appropriate weight in the consensual decision which follows.

Note that while aligned with the Thomist school of philosophical thought broadly referred to as "proportionalism", *proportionism* as we understand it, is not tied to Aristotelian natural law theory as tightly as proportionalism. Regarding the latter, Aquinas argues that natural law is absolute but that, 'in certain matters of detail', natural law needs to be guided by reason (Aquinas 1952 [Part 1 of 2nd Part, Q94:4]). For example, proportionalists might say that sometimes a greater good can justify contravening the law. That is, if a man was starving, it would be permissible to steal rather than starve to death; and if a mother's health was threatened by pregnancy, the pregnancy could be terminated. Sometimes, this is reduced to the principle of lesser evil. For the proportionalist, the rule itself is absolute but it can be amended by circumstances. The allied principle of 'double effect' might be re-titled by the sceptic as 'double think'. As we understand and use the term 'proportionism' however, the mean or balance point for a final decision (the 'rule in this instance', as it were) is determined by the process of dialogic consensus itself. This is a fine but important distinction, in our view.

The modern development of the Proportionist approach, as we offer it, can be found in part in Joseph Fletcher's book, *Situation Ethics*, in which he offers a method of situational or contextual-based moral decision making. Fletcher argued that a *framework* of ethics was not possible, only a *method* by which to approach moral decision making. He proposed that 'in actual problems of conscience the situational variables are to be weighed as heavily as the normative … constants' (Fletcher 1966). Adherence merely to rules about permissibility and impermissibility effectively removes conscience and indeed the moral agents themselves from the decision-making process. In situation ethics, the moral agent is the decision-maker, judging what is best in the particular circumstances and allowing for the foreseeable consequences. His method is based upon a sophisticated love of neighbour *as* oneself—described by Fletcher as an 'agapeic calculus' (Fletcher 1966). Other moral principles, maxims, rules, and guidelines serve simply as 'illuminators … not directors' (Fletcher 1966). It is in the emphasis on the role of the moral agent that we see the connection with virtue ethics, as inspired by Aquinas, and it is this emphasis that we have tried to capture in the notion of proportionism, similar but philosophically distinct from proportionalism, as identified above.

Charles Curran argues for what he describes as a theory of compromise. From a theological perspective, he apportions natural law into primary and secondary natural law. Primary natural law is the state of human existence before the fall of the world into sin. Secondary natural law is the state of human existence after the fall into sin. He also distinguishes absolute natural law from relative natural law. Absolute natural law is based on the 'ontological, abstract human nature' (Curran 1979). Relative natural law is based on the actual reality of the human situation. In individual morally dilemmatic situations, the nature of absolute natural law is unchanged. However the abstraction of absolute natural law is applied differently in different situations—'the formal demands of the absolute natural law remain the same, but they are abstractions which are then applied differently in different situations' (Curran 1979). Examples of situations which fall under relative natural law rather than absolute natural law include killing in self-defense, just war, and capital punishment, amongst others (Curran 1979). Wrongness is not exclusively in the act itself so much as in the situation within which the act is forced to be done. 'From one point of view the action is good, because it is the best that one can do. From another viewpoint the action is wrong; that is, it manifests the sinfulness of the situation' (McCormick 1967).

We contend that the most apposite approach to moral decision making in clinical practice is set in the context of the particular patient. A greatest Good, utilitarian assessment of consequences needs to be balanced by a proportional awareness of fundamental a priori rules; and a rules-based decision needs to be balanced by an appropriate proportion of empirical awareness of the situation the patient is in. Jürgen Habermas agrees when he writes that, in clinical contexts and considering the doctor-patient relationship, there must be a 'contextual sensitivity and prudence on the one hand and autonomy and self-governance on the other' (Habermas 1990).

Two clarifications then follow (Walker and Lovat 2016). First, the substantive frameworks and principles are clearly of value in moral decision making. It is a question of balancing them, seeking to maximise the virtue in the decision to be made. Sometimes the balance will lean towards deontological rules, and sometimes towards teleological utility. Sometimes the balance will lean towards autonomy, and sometimes towards beneficence. Second, the proportionist approach, aware of the context of the clinical decision, and sensitive to the need for dialogue amongst those affected by the decision, is not saying that morality is simply ethical relativism. The frameworks of deontology and teleology set the limits or boundaries to the box within which the dialogue is to be had, and hence within which the decision is to be made. For example, in the case of a baby potentially born with anencephaly, the deontological maxim states that it is impermissible to actively kill the baby at birth. Dialogue about moral options begins from this maxim.

A proportionist approach allows for tolerance of anomalous and/or conflicting positions in an ethical dilemma where interpretations offered by both the deontological and the teleological frameworks are valid, but both need to be moderated and made complete by an empathic compassionate caring, self-insightful and wise clinician in communicative discourse with the participants in the dilemma. Thus, together and in the Greek, they achieve *synderesis* (practical wisdom) in order to impel *praxis* (practical action) which results in the *eudaimonia* (flourishing) of all in the dialogue.

As if to tie together the threads of phenomenology and proportionism, Fried writes that a person:

> has a right when he is confronted by another in a concrete situation to demand that his particular situation be taken into account … the professional who undertakes to deal with a patient's serious illness by that undertaking is obliged not only to acknowledge but to respect, to make provisions for the peculiarities, the needs and values of that individual (Fried 1974).

To our minds, understanding that there is no independent right-making property of a moral decision, *demands* a balanced or proportionist perspective.

Habermas writes that '[n]either interrogations nor analytic conversations between doctor and patient may be considered to be discourses' (Habermas 2001) in the sense of a cooperative search for truth in the clinical situation at hand. A dialogical approach to moral decision making is preferable, especially in clinical situations. This approach will be considerably expanded in Chaps. 4 and 5. In the terminology of models for the doctor-patient relationship, it most closely fits with the *shared decision-making* model (Hoffman et al. 2014). It is predicated upon exploring and respecting "what matters most" to the individual patient— Pellegrino's perceptual good.

3.3 Summary

As noted in Chap. 2, in the secular Western tradition, three frameworks are traditionally recognized as offering guidance for ethical decision making. Maximizing the Good of the patient is the final purpose of medicine. In this chapter, four action-guiding principles distilled from the normative frameworks have further helped to guide ethical decision making in clinical situations. Especially in our current epoch of widely disparate value systems, both patients and clinicians may bring widely disparate perspectives to the consultation. In order to maximize the good of the patient, clinicians should seek a balance between a priori rules and empirical consequences. This approach is framed here as proportionism. It can be put into practice, seeking a virtuous mean amongst ethical imperatives as viewed from different perspectives, via a dialogue amongst those involved in the decision to be made. It is to the principles which underlie this dialogue, to which we now turn our attention. It is our aim to move from what could be termed the motherhood statement, "we need to talk with our patients", in order to fortify the foundations of the dialogue we need to have by arguing for its moral philosophical basis.

References

Aquinas, T. 1952. *Summa theologica*. (Fathers of the English Dominican Province, and D. J. Sullivan, Trans.). In *Thomas Aquinas II* vol. 20, 1st ed. Great Books of the Western World. Chicago: William Benton, Encyclopaedia Britannica Inc.

Aristotle, 2000. *Politics*. Translated by B. Jowett, Revised ed, ed. Mineola, New York: Dover Thrift Editions.

Beauchamp, T.L., and J.F. Childress. 2013. *Principles of biomedical ethics*, 7th ed. New York, Oxford: Oxford University Press.

Carel, H. 2011. Phenomenology and its application in medicine. *Theoretical Medical Bioethics* 32 (1): 33–46. doi:10.1007/s11017-010-9161-x.

Clouser, K.D., and B. Gert. 1990. A critique of principlism. *Journal of Medicine and Philosophy* 15 (2): 219–236.

Curran, C.E. 1979. Transition and tradition in moral theology. In *London*, xiv, 255. University of Notre Dame Press.

de Beauvoir, S. 2011. *The second sex*. (Constance Borde, and S.M. Chevallier, Trans.). New York: Random House.

Deligiorgi, K. 2012. *The many faces of Kantian autonomy*, 1st ed, The scope of autonomy: Kant and the morality of freedom. Oxford: Oxford university Press.

Fletcher, J. 1966. *Situation ethics: The new morality*. London: SCM Press.

Fried, C. 1974. *Medical experimentation: Personal integrity and social policy* Clinical studies series, vol. 5. Amsterdam: North-Holland Publishing.

Habermas, J. 1990. *Moral consciousness and communicative action*. (C.L.S. Weber, Trans.). Cambridge: Polity Press.

Habermas, J. 2001. *Justification and application: Remarks on discourse ethics*. (C. Cronin, Trans.). Cambridge: MIT Press.

Harris, J. 2003. In praise of unprincipled ethics. *Journal of Medical Ethics* 29 (5): 303–306.

Herring, J., and P.-L. Chau. 2007. My body, your body, our bodies. *Medical Law Review* 15 (1): 34–61.

Hoffman, T.C., F. Legare, M.B. Simmons, K. McNamara, K. McCaffery, L.J. Trevena, et al. 2014. Shared decision making: What do clinicians need to know and why should they bother? *Medical Journal of Australia* 201 (1): 35–39.

Jones James, and The Tuskegee Institute. 1981, 1993. *Bad blood: The Tuskegee syphilis experiment*, Revised ed. New York: Free Press.

Larmore, C. 2008. *The autonomy of morality*. Cambridge: Cambridge University Press.

McCormick, R.A. 1967. Notes on moral theology: January–June, 1967. *Theological Studies* 28 (4): 749–800.

Mill, J.S. 1952. On liberty. In *American state papers, The federalist, On liberty, Representative government, Utilitarianism*. Great Books of the Western World, vol. 43, 1st ed., 267–326. Chicago: William Benton, Encyclopaedia Britannica Inc.

Newson, A. Genomic advances and testing and screening before birth: What's at stake? In Australian Association of Bioethics and Health Law Conference, Sydney, 14 July 2013.

Pellegrino, E. 2013. Moral choice, the good of the patient, and the patient's good. In *Ethics and critical care medicine*, ed. J.C. Moskop, and L.M. Kopelman, Philosophy and Medicine, vol. 19. xx, 236. Dordrecht: D Reidel Publishing.

Raanan, G. 2003. Ethics needs principles—four can encompass the rest—and respect for autonomy should be "first among equals". *Journal of Medical Ethics* 29 (5): 307–312.

Walker, P., and T. Lovat. 2014. Concepts of personhood and autonomy as they apply to end-of-life decisions in intensive care. *Medicine, Health Care and Philosophy* 18 (3): 309–315. doi:10.1007/s11019-014-9604-7.

Walker, P., and T. Lovat. 2016. Towards a proportionate approach to moral decision making in medicine. *Ethics and Medicine* 32 (3): 153–161.

Walker, P., T. Lovat, J. Leitch, and P. Saul. 2014. The moral philosophical challenges posed by fully implantable permanent pacemakers. *Ethics and Medicine* 30 (3): 157–165.

Weindling, P. 2001. The origins of informed consent: the International Scientific Commission on medical war crimes, and the Nurenberg code. *Bulletin of the History of Medicine* 75 (1): 37–71.

Wolpe, P.R. 1998. The triumph of autonomy in American medical ethics: A sociological view. In *Bioethics and society: Constructing the ethical enterprise* ed. Raymond DeVries, and J. Subedi, xxiii, 276. New Jersey: Prentice Hall.

Chapter 4
The Foundations and Benefits of Dialogic Consensus

4.1 So What Is Different About Our Contemporary Era?

The contemporary era in which we live and work is termed in this book *Post-modern*. It can trace its origins to the completion of post-World War II reconstruction. The post-modern era is characterised by rapidly increased technology throughout the world, which is important from a moral philosophical perspective, in at least two ways.

The first is that the moment of moral inter-connectivity is based upon personal interaction. If one bumps into another while walking, it is easy to offer an immediate apology, which is likely to be accepted. If a car driver infringes upon another, then a rude gesture or horn blast is likely to follow. The further we are disconnected from each other, especially via technology, the less easily we relate to each other as moral beings. Consider weapons. A fist fight requires close eye-to-eye contact. A knife, sword, and bow and arrow, progressively distance and thus disconnect, the protagonists. This is made worse by hand-guns, rifles, artillery, rockets and drone aeroplanes which can target enemies out of sight over the horizon. Because clinical encounters are seen by us primarily as inter-relations amongst persons, they are necessarily moral encounters.

The second is that technology has especially impacted upon global communication—via television, the internet, and social media. These have spread a much wider knowledge of different cultures, ethical values, ways of living, and personal opinions. Through widespread travel and immigration, people from a variety of different cultures and belief systems have been brought into our communities. The resultant socio-cultural, ethno-national and religious diversity results in potentially conflicting life views and value constructs. These diverse values cannot readily be arranged into a hierarchy, especially without considering their context. Accompanying these developments is a considerable diversity of moral values and hence moral pluralism, which is the "elephant in the room" during clinical consultations.

© The Author(s) 2017
P. Walker and T. Lovat, *Life and Death Decisions in the Clinical Setting*,
SpringerBriefs in Ethics, DOI 10.1007/978-981-10-4301-7_4

We suggest that contemporary society should, properly, be viewed as a network of inter-relations amongst people, and amongst their values and meanings. Values and meanings which can only be shared and be understood, via language. We believe that it is not an overstatement to say that, contemporaneously, language 'constitutes, rather than reflects, the world' (Bertens 1995). It is language which impels knowledge and understanding. At a simple level, being told that "you have AIDS" profoundly reorientates the patient's lifeworld and (generally) impels significant change, based upon the perceived sequelae associated with those words. On a broader level, since society and societal norms are constantly shifting, this results in language and meaning constantly shifting. In the same way that the prognosis which follows upon the word "AIDS" is improving, this shifting within language results in the meanings attributed to Right and Good, being recognisable at a multiplicity of depths, to different members of pluralist society.

This also means that, in our contemporary era, we should be aware of multiplicity of truths, and so de-emphasize any single ethical framework as universally applicable independent of context. Hence we favour a *process* over a substantive framework, emphasising community consensus rather than subjectivity for its action-guiding normative force. Using communicative language as the construct, we seek to determine what is morally good, right, and just, in the situation at hand.

In philosophical terms, we argue that clinicians should recognise their responsibility to otherness, open to difference, dissonance and ambiguity. Post-modern ethics incorporates a strong '"injunction to listen" to the other, a willingness to hold open an intersubjective space in which difference can unfold in its particularity' (White 1991). Medically, since our post-modern epoch is characterised by an ever-increasing armamentarium of life-sustaining technology, an active process of moral decision making in clinical situations, rather than mere contemplation, is required.

The introductory chapter proposed that moral philosophy be informed by the question "how should I act?", or better, the question of Socrates, "how should we live?" Understanding that each clinical doctor-patient contact does indeed have a basis in moral philosophy, impels clinicians to move their perspective from themselves exclusively or predominantly (*ego*) towards others as well (*alterity*). The concept of alterity captures an awareness of others, our necessary inter-connectedness, and that fact that alternative viewpoints to one's own, exist and are likely to be just as valid as our own (Levinas 1999). Thus, the question of Socrates could now be re-conceptualised as "how should we live, together?" Moral decision making are properly derived from a process of inclusive, non-coercive and self-reflective dialogue within the community affected.

Many clinicians will be comfortable with the scientific method of empirical proof for the propositions we make. There are, however, no absolute moral tags or identifiers which say that one or other action in an ethical or moral dilemma is "clearly" right or correct. That is, there are no moral facts. Further, we cannot design an experiment which will prove the rightness or wrongness, correctness or incorrectness, permissibility or impermissibility, or the truth or falsehood of an action we might take in an ethically or morally dilemmatic situation. Empirically we can scientifically prove whether it rained yesterday, or prove that gravitational fields exist. Moral judgements cannot be subjected to the truth conditions of empirical or scientific experimental proof. Put another way, 'morality is an independent domain of thought' (Dworkin 2011). If I want the right to hold that a moral contention is true, in the sense of an action being permissible or impermissible, then I have to provide arguments in support of that contention. Put bluntly, '[t]here is just no other way' (Dworkin 2011). The moral arguments we provide must have a basis in reality, must be meaningful, and must be aware of the context in which the moral decision is being made. This can only occur if they are developed in the situation of a dialogue or discourse about the dilemma at hand. If you change your ethical stance, then you can only have been persuaded to your new stance by arguments—cogent, relevant arguments, which will often involve exploration and consideration of the values which are important to the participants in the dialogue, and which you listened to.

Thus we return again to the importance of language. We argue that language is the necessary intermediary for moral dialogue. Language relates to truth, meaning, and value. It is not possible to have a discussion about morals if we do not have the words to articulate the concepts, and if we do not agree about the meaning of the words. If the dialogue is to be unforced, then participants need to understand how the use of language can result in a coerced agreement rather than consensual agreement. Recall that in Chap. 1 we argued that ethics is a more individual subjective assessment of values as relatively good or bad, while morals is a more collective and intersubjective assessment of what is right for all those affected. Hence, decision making in the clinical encounter should be approached from the perspective of moral dialogue, rather than ethical monologue. This is more than a pragmatic (or strategic) *modus vivendi* or "we agree to disagree", and means that this morality must have a binding character which transcends competing value conceptions (Forst 2014). In practical application, dialogic consensus relocates decision making away from what may once have been a clinician-dominated monological reflection upon imperatives, utility, or an agapeistic (loving) calculus, or upon the derived principles of autonomy, beneficence, non-maleficence and justice. The moral virtue (goodness or rightness) of a clinical decision cannot be abstracted from the decision's context. Similarly, neither clinicians nor patients are independent from their inter-personal relationships and historical-socio-cultural backgrounds. Appeal to the normative frameworks or to principles derived from them is therefore less and less adequate as a practical way of deciding on appropriate moral action—especially in clinical situations. Dialogic consensus relocates decision making into a social space cognizant of the other, wherein all involved—patients, relatives, clinicians and other staff—engage in a conversation, a process of discussion or dialogue, of evaluation and re-evaluation, in order to explore the values held by those affected by the decision, with the aim to achieve a consensual understanding of what is important to each participant.

Note however, that we are not at all saying here that ethical questions are less important than moral questions. Self-reflective critical knowing necessarily evaluates answers to complex questions about how I might live my own life, in the context of my lived experiences, motivations, and my final purpose as I perceive it. *Our* approach to answering moral dilemmas for *us*, however, is different to how *I* monologically approach ethical dilemmas for *me*. Taking seriously our responsibility to make ethically good or just decisions for ourselves and our actions, is a fundamental part of living a good life. In that sense it implies sensitivity and respect for others. This awareness of others, and of our shared humanity, means that we can say that ethical decision making promotes moral decision making. Put in more philosophical terms, moral decision making is also procedurally-distinct from ethical decision making (Forst 2014). That is, ethical norms are a priori (prior to argumentative discourse), while moral norms are a posteriori (after argumentative discourse). Moral decision making should not be conflated with ethical decision making.

That said, now however we need to progress from the "motherhood" statement that we need to have a conversation, a dialogue, with our patients. We need to propose principles which fortify the foundations of dialogic consensus following an inclusive and non-coercive reflective discourse. This understanding of dialectic is especially apposite in coming to understand how we can best maximise the good of our particular patient in their particular situation. Furthermore, we argue that the process of dialectic, leading to consensus, has of itself a distinct moral dimension. Similar to such 'assisted conversations' (Fiester 2015) is what has become known as bioethical mediation. In seeking to fortify the foundations of dialogue, we favour two principles of Jürgen Habermas. The two principles we look to are his discourse theory of morality and his principle of communicative action. Both recognise our essential inter-connectedness; related in turn to intersubjectivity—by which we mean 'an intermeshing of the perspective of each with the perspectives of all' (Habermas 2001a).

4.2 The Discourse Theory of Morality and Communicative Action

Habermas captured Kant's principle of the universalizability of his categorical imperative and widened its social applicability by reformulating it in his discourse theory of morality as requiring that *all* affected people must be able to agree that it is universalizable. All participants need to agree that the decision which we make, can be universalised to all who are affected. While individuals can of course have moral thoughts, an isolated individual cannot, monologically, determine a moral norm which is applicable to others. Moral thoughts have no normative force upon others unless all in a community, after public communal dialogue, agree. Kant's generalizability criterion gestures towards incorporating others, but Habermas' discourse theory of morality embodies it. The dialogue which follows is predicated upon principles of communicative action.

In communicative action, speech acts are orientated to understanding. For Habermas, consensus via mutual understanding is the final purpose of language, and it lays the foundation for a normative relationship (a sense of oughtness or shouldness). He terms the alternative to communicative action to be strategic action. In strategic action, speech acts are orientated to success, aiming to influence, and are associated with power. He may agree that strategic action characterises oratory, while communicative action characterises discussion or dialogue. Specific to the language of clinical decision making, it is very common to refer to "clinical judgements" versus "patient or parent wishes" (McDougall and Gillam 2014). This language of the "judgement" of clinicians implies expertise, rationality, and validity. It privileges the clinician's views as being reasoned, whereas, the "wishes" of the patient or parent imply mere preference and so lack validity. Making a moral judgement requires collecting the facts, and then reasoning about them based upon a set of ethical values. Both clinicians and patients/parents are able to make moral "judgements", from the value set they have, respectively, as "good clinicians" and as "good parents". It is incorrect to portray the views of the clinicians and the parents, when they conflict, as different in validity.

In his theory of communicative action, Habermas aims towards truth-seeking via participatory democracy. He argues that the use of language, in the sense of either linguistic, or non-verbal, communication, aims 'to attain consensus in a context in which all participants are free to contribute and have equal opportunities to do so'. It has been characterised as a form of linguistic interaction 'where all speech acts contain validity claims concerning comprehensibility, sincerity, truth and justification, which are openly criticizable and discursively redeemable' (Scambler 2001).

4.3   Dialogue, Consensus, and Dialogic Consensus

4.3.1   Dialogue

In common usage, the words dialogue, discussion, discourse and the populist "have a conversation", are often interchangeable (Walker and Lovat 2017, In Press). The more formal notion of *discourse* has various meanings, depending on the academic discipline and situation. Discourse, in general, refers to the use of language as a part of a social practice. Discourse may also refer to established ways of constructing the meanings of phenomena, knowledge and reality, and the language which links them. The emphasis is on the elucidation of knowledge in terms of its meanings and values. Conceptually, it is a systematically organised body of linguistic and other meanings as it is embedded in social relationships and cultural practices. A discursive approach enables one to explore the construction of meanings in human interaction.

In this book, dialectic (from the Greek *dialegesthai*, to converse, and *dialegein*, to sort or distinguish) means 'to pass from one part—an object, a notion, a problem—to another by the means of language and reason' (UNESCO 2007). The association of dialectic with truth-seeking after reasoned argument, amongst people with different perspectives or points of view, has a very long history. In _The Sophist_, Plato's Socrates contrasted dialectic with sophistry. Sophists were paid orators who sought to argue the case they were assigned, without regard for the truth of their argument. Philosophers, on the other hand, favoured dialectic—in which they offered and received arguments, evaluated them for truth and meaning, and thus sought to discover truth in the arguments presented. Importantly, individuals in the dialogue may need to change their views, in order to arrive at the truth. Since actual people are in a clinical situation, actual people need to dialogue in order to determine what matters to them, and so clinician and patient, together, should make the morally best decision in the situation in which they actually find themselves.

Conceptually, it is a systematically organised body of linguistic and other meanings as it is embedded in social relationships and cultural practices. Since actual people are in a clinical situation, actual people need to dialogue in order to determine what matters to them, and so clinician and patient, together, should make the morally best decision in the situation in which they actually find themselves.

During a clinical consultation or a case conference, participants must aim to foster ideal speech conditions for a dialogue. Each speaker speaks truthfully and non-coercively, rather than aiming at influencing or coercing other participants in the dialogue. The aim is to reach a consensus in the decision, via mutual under-standing. Seeking truth in the outcome requires participants to be aware of "ways" of knowing. Habermas describes three "ways" of knowing (Habermas 1972). The first is collection of the data or facts about the matter (empirical-analytic knowing). The second is coming to understand the connections amongst the facts, including any prior beliefs and heritage impacting on meaning (historical-hermeneutic knowing). The third is critically reflecting upon the facts and their meanings in the context of our own self, the one who is aiming to know (self-reflective, critical knowing). Empirical-analytic knowing is most straightforward, although, as already noted, that is not to be taken to mean that this book views all medical outcomes as quantifiable with certainty. There will always be a level of uncertainty, often sig-nificant uncertainty, about what outcome an individual patient will have following treatment. Historical-hermeneutic knowing follows upon understanding the importance of meanings and values. In other words, there must be an attempt by all participants to understand the actual reality of the patient's situation, and how this affects the goods of the patient and their family; for example, what is actually important to them? Thus, the patient's embodiment both in their situation, and in-relationship with those others around them, is recognised. Self-reflective critical knowing is driven by our interest in being emancipated in our knowing, being freed from unhelpful preconceptions, and outdated or incomplete beliefs. Achieving dialogue at this level is necessary for the practical outcome of dialogic consensus.

The assumptions which necessarily underlie the dialogue are important to the argument we are making (Walker and Lovat 2016) These assumptions dictate that each participant mutually considers each other to be accountable; and they mutually consider each other ready and willing to reach mutual understanding. That is, each acts so as to aim to reach consensus (Habermas 2001b). In other words, in practice participants use language in the same way, all relevant arguments are brought to the dialogue, each is allowed to participate and express their attitudes, wishes and needs, each can introduce or question any proposal, there should be no compulsion applied by or toward to any speaker (Jones 2001), and each participant genuinely tries to understand the perspective of the others. In ideal speech situations of undistorted communication, concern is only for the most valid argument—'the unforced force of the better argument' (Habermas 1996). Language underpins meaningful dialogue and is something which we, all of us in the dialogue, must use and understand in the same way. In a clinical setting, it is of fundamental importance that (for example) the meaning of medical terminology is explained clearly to all the participants. But equally importantly, other language needs to be understood. For example, "do everything you can", may mean to the speaker, "but don't allow needless suffering" or "unless the outcome is likely to be futile", and these meanings must be explored. Discourse is thus used here in its technical sense as linguistic blocks of contextually-situated 'written and spoken language produced as part of the interaction between speaker and hearers and writers and readers' (Candlin et al. 1999). Ultimately, language and understanding meanings, precedes consideration of competing truth claims; which are themselves a feature characteristic of the pluralism extant in contemporary society.

Bent Flyvbjerg enumerates practical requirements for 'validity and truth' in a discourse, in greater detail. These requirements are: (1) generality—by which he means that no affected party should be excluded from the discourse; (2) autonomy—participants have equal possibility to present and criticize validity claims; (3) ideal role taking—participants are willing and able to empathize with each other's validity claims; (4) power neutrality—existing power differences between participants, especially that power vested in clinicians, are to be neutralized so they have no detrimental impact upon consensus; (5) transparency—participants must openly explain their goals and intentions and avoid strategic action; and (6) sufficient time (Flyvbjerg 2000). As well, there needs to be an awareness that in medical consultations there is usually a considerable emotional flux present during the dialogue. Each participant, in principle, has an equal right to solve problems, equal duties, and equal co-responsibilities. Karl-Otto Apel argues that this is the 'ideal situation which we *must* anticipate when entering into the discourse' (Griffioen and Van Woudenberg 1990).

Through self-insight and self-reflection, members recognise that each participant brings to the dialogue their own historical and socio-cultural background, and each needs to be aware of this in themselves and in others. In the context of a clinical case-conference, the aim is to establish a situation of non-coercive dialogue, so that a consensus can be reached. The participants each seek to present and to understand the facts, and the meanings that each attaches to those facts. This is the only way to know the good of the patient in order that it can be maximised. For the clinician who is taking self-reflective knowing seriously, there is a strong empathic compassion for the other as a suffering vulnerable patient, rather than being there as an authority figure. They may usefully be aware of quantitative research which (for example) shows that in multidisciplinary groups in Danish Hospitals, compared with nursing staff, physicians use more of the discussion time, use a more assertive style of argumentation, and the solutions chosen are usually first proposed by physicians (Holm et al. 1996).

Habermas stresses that communicative action is not identical with communication—although it takes place by means of communication (Outhwaite 1994). He describes it as a type of interaction coordinated through speech acts but not coincident with speech acts. What then of non-verbal communication—for example aiming to reach consensus about whether to offer a cochlear implant to the child whose parents are both deaf? The process of communicative action is more difficult, both because the educational achievement of profoundly deaf adults may be less than hearing adults, which, conceivably, may limit their capacity to discuss relevant concepts, and also because of limitations around the need to write down questions and answers or use a deaf interpreter to relay via signing. Therefore, it may be that the dialogue exists at a lesser level of sophistication and so has less normative force. Habermas might agree that communicative action in this setting simply requires more time and effort from the participants, and that deaf parents are uniquely able to completely understand the experience of deafness, in a way hearing parents (and hearing clinicians) are not able to. Having experienced deafness themselves, deaf parents do not have the initial fears hearing parents often have, and so can more realistically weigh up the pros and cons of interventions such as cochlear implants. Non-verbal communication is often enhanced in the hearing impaired, which may indeed avoid misunderstandings related to the limitations of language.

In the dialogue of a medical case conference, each mature participant brings some moral sensibilities, non-interrogated though they might be. As already noted in Sect. 3.2, it may be that the three normative ethical frameworks, or the four principles derived from them, or professional codes of ethics, world ethical declarations, and similar position statements, set the boundaries of the box within which the dialogue begins. Borrowing from Susan Wolf, it may be that in the dialogue, an individual may not be able to follow their preferred moral principle, but will recognise that another principle, agreed to by all, may not, rationally, be unreasonable in the context at hand. During the dialogue, 'the recognition that everyone rationally *could* accept a principle may count as a reason for someone *to* accept the principle' (Wolf 2011). Practical ways of reconsidering dissensus will be discussed further, below in Chap. 5. It is the contention of this book that moral truth is attained through consensus, on the basis of conditions set out above, and this is the basis for normative force.

In practical terms common sense determines how many need to be actively involved in a particular case, and also what relative contribution might be judged as appropriate. Consider a case conference after a serious head injury. Weighting of the prognostic guide offered by the neurosurgeon would intuitively be relatively more important because s/he would seem most likely to have the entitlement to prognosticate about the likely outcome from this particular head injury. For the same reason, listening closely to the relative who is offering their knowledge of what the patient would want is also relatively more important. Input from the discharge planner as to what facilities for rehabilitation are available near-by or distant is relatively important. The business manager might be listened to in terms of costs but ideally would not try to predict likely neurological outcome from this particular injury. Apel argues that the community, in principle, is infinite—no-one can be excluded without a reason. Pragmatically, nonetheless, many must be excluded in order to avoid an utterly unwieldy dialogue. Some restrictive conditions are a practical necessity.

4.3.2 Consensus

Second, the word *consensus*. As we have argued elsewhere there are important differences amongst the words unanimity, acquiescence, and consensus (Walker and Lovat 2016). Unanimity is agreement by all participants, both publicly and privately. Acquiescence is agreement out of a sense of benevolence, of altruism, of coercion, or another reason which denies true argumentation. Consensus is general agreement, following argumentation, in reaching a decision about what is best for the group or the community which is making the decision. Hence, and importantly, some of the participants could disagree with the decision itself, but still agree that it is the best decision for the group as a whole. Put another way, it may be possible for participants to accept a position which it is not reasonable for them to reject, and so reach consensus. Individual dissensus is not fatal to moral decision making (Wilkinson et al. 2015). That is, a range of reasonable (or not unreasonable) decisions may prove acceptable—recall the "zone of parental discretion" noted above in Sect. 2.5. As well, we argue that the process of moral decision making in clinical settings itself provides normativity, rather than implying the absolute truth of the decision which is reached.

Nor is consensus simply a vote, where the majority decision is chosen as the morally appropriate path. By itself, agreement by vote does not imply moral truth. A recent clinical ethics paper reported that the clinical ethical decisions made in a hospital-based study were thought to be "mostly right" (Doran et al. 2015). This assessment appears to have been based upon an essentially unexamined, majority decision, made predominantly by clinicians, and based on the "settled morality" which is "part of the fabric" of the hospital setting and articulated in hospital policies and guidelines which regulate interactions between clinicians and patients (Doran et al. 2015).

Consensus links its moral authority to the consent of the participants in the dialogue (Jennings 1991). Consensus and consent have the same etymological root (Caws 1991)—deriving from the Latin *consentio*, for 'to feel together, to agree'. However, as noted, consent to an outcome reached by vote, is not consensus. Nor is agreement by trading-off some other agreement in exchange. And nor is pragmatic agreement, in order to reach a decision, especially by a certain time. These may be recognised as compromise, as a way of reaching a conclusion, rather than consensus (Caws 1991).

Consensus is tolerant of value pluralism. It is aware that different groups within society recognise different conceptions of the *good*, each of which may be viable, nor arranged in a definite hierarchy. The inevitable differences which follow can result in conflict. Hence, tolerance is an essential corollary to pluralism. Given this, only genuine, mutually respectful and transparent discussion is likely to resolve moral conflicts (Kerridge et al. 2013). In the context of dialogue in actual concrete situations, values may need to be ranked, affirming that all perceived goods have value, but ultimately, some subset of values must actually be agreed-to by the participants. Those chosen may include values to which we ourselves do not subscribe. An obvious clinical example is that of an adult Jehovah's Witness who refuses blood transfusion, even to avoid death, because of the belief that it will result in the loss of their eternal soul. The import of incommensurability means there should be pragmatism and, ideally, flexibility in choosing amongst different goods by the participants in the dialogue, remaining attentive to the specifics of the situation while aware of the claims and the circumstances of those affected by the decision (Crowder 2003). Consensus implies respect for 'the full range of human goods and lives, including those we cannot accommodate within our own decisions' (Crowder 2003).

What has become known as "shared decision-making", wherein clinicians and patients make decisions about treatment in partnership, is not only soundly based upon moral philosophical principles, but has been shown to reduce mortality, reduce readmission rates, reduce healthcare acquired infections, reduce length of stay, enhance compliance, and improve functional status (Australian Commission on Safety and Quality in Health Care September 2011). It is important that we recognise that moral dilemmas in clinical settings (but not limited to those settings) may be very complex. Not the least reason for this is the rapidly advancing pace of medical technologies, which are able to save or preserve life in ways not previously considered (for example, fully implanted pacemakers, discussed in Sect. 3.1.1). As already noted, a simple solution is rarely possible; and rarely should it be expected. The fact that the process of moral decision making may be difficult, simply underlines the seriousness of the decision being made, and is not fatal to the process.

An additional advantage of this process accrues in terms of the psychological well-being of the parents. It is common for parents to ask, during a case conference or consultation, the anguished question "how can I make this decision?" The rise and rise of patient autonomy as a dominant principle of medical decision making means the response is traditionally along the lines of "you must" or "only you can make this decision". Understanding the process of the discourse theory of morality and of communicative action, offers the singular advantage to the parents that they can be reassured that they don't have to make the decision alone—that "we", those in the community having the dialogue, make the decision jointly and that decision thus has normative force. From a virtue ethical compassionate and caring perspective, the family is then less likely to have lingering unresolvable doubts about whether the normatively right decision was made, at a later stage, if this concept of normative force by way of the *process* undertaken, is understood and reflected upon. This outcome, that of compassionate reassurance for the parents that the "right" or "best" decision was made, contributes greatly to their good as parents and members of a family.

There is also an advantage for clinicians, in that they can come to know that even previously unmet morally-challenging situations can be approached successfully by following this process.

4.3.3 Dialogic Consensus

The historical reliance of medical ethics on substantive frameworks and the principles derived from them, is no longer viewed as sufficiently aware of moral pluralism to remain useful in our current era (Walker and Lovat 2017, In Press). In an era characterized by post-modern understandings of moral decision making, the clinician recognises the essential inter-relationship of persons, and, further, that this is based on meanings—that is, on language. A properly constituted dialogue considers the context of the situation and the values which the participants hold. It evaluates these seeking a reflective equilibrium through communicative action with others in the moral situation—which then impels a consensual decision. We argue that it is only through dialogue and dialogic consensus that we can ground an approach to morality which is appropriate to clinical decision making in our times.

4.4 An Example of a Case Conference Using Dialogic Consensus

As an illustrative case study for dialogic consensus (painfully aware that any example is vulnerable to criticism), consider Baby 'H'. Baby 'H' has had an in utero MRI diagnosis of Congenital High Airway Obstruction Syndrome due to tracheal atresia (absent windpipe) at 22 weeks gestation. The decision is whether to offer an EX-utero InTra-partum (EXIT) procedure (Walker et al. 2005) wherein an airway is established while on placental support, while the baby has only the head, neck and one or both shoulders delivered from within the uterus. Tracheostomy must follow, and then multi-staged tracheal reconstructions. Mum has gone through four years of in vitro fertilisation (IVF) and, at 44 years old, this is arguably her last chance to conceive. The family have an older child, aged six, who is perfectly well. Without successfully establishing an airway at birth, Baby 'H' will not survive.

A case conference is formed. The participants are informed of the process and their co-equal responsibilities in the process. An interpreter is not required. The participants are limited to the parents (at their request neither the grandparents nor the six year old sibling are involved), the paediatric otolaryngologist who will establish the airway, the paediatric and maternal anaesthetists, and the neonatal intensivist. The parents accepted the offer of participation from both their family priest and the NICU social worker, who acts as the facilitator.

Deontological precepts might be introduced and acknowledged via articulating the special joy of a baby within, being aware of the essential personhood and inherent dignity of a human baby, independent of function or contribution to society, who cannot permissibly be deliberately killed by the doctor (but who can permissibly be allowed to be born, and die at birth).

Teleological consequences include articulating in a way which encourages all in the dialogue to contribute, difficulties in terms of risks to the mother during EXIT procedure, care of a tracheostomy in a new-born, difficult multi-staged recon-struction of the trachea, which may cause suffering to the baby, and may be unsuccessful. Normal speech and swallowing cannot be assumed. There will be significant time off work for both parents, with corresponding financial conse-quences, consequences to the family dynamics in terms of the older child, and the potential for parental divorce. A study of 'severely unhealthy infants' reviewed 12–18 months after birth reported a ten per cent rate for the parents no longer cohabiting (Reichman et al. 2004). There will also be direct financial and missed opportunity costs to society in general, including the potential abandonment of a severely hypoxic baby.

The vision of a virtue ethical framework, maintaining empathy, compassion, and caring, striving indeed for wisdom, overlies the dialogue. During the case confer-ence participants explore their understandings of the knowable facts about the condition and range of prognoses for baby 'H', their own values, and their desire to maximise all the Goods of the Baby 'H' and her family. Despite the likely disparate social and moral backgrounds of the participants, during the medical case confer-ence for Baby 'H', good clinicians and good parents dialogue about the morally conflicted situation they are in. After dialogue based upon principles of commu-nicative action, in which all participate, the consensus in the decision may be that notwithstanding the human dignity of Baby 'H', the likely burdens of life for the baby 'H' and her family, do not outweigh the potential benefits. As well, the family believe that their child will go to heaven. Thus a proportionist approach reasons and reflects that to intervene with a surgical tracheostomy at the moment of birth not an appropriate action for Baby 'H'. Thus the consensual decision which the case conference comes to, is that we provide compassionate support for the parents and the sibling as the pregnancy continues, but that we do not intervene surgically to prolong life at birth.

The awareness that, when one takes seriously each individual's perspective, there is always a multiplicity of truths, grounded in the context at hand and informed by participants' socio-cultural and/or religious beliefs, must impel clini-cians to dialogue with their patients and family in order to achieve consensus in clinical decision making. While allowing for the reality that achieving absolute truth might be beyond any human process, the process described here seeks the best possible outcome in the circumstances (Walker and Lovat 2016).

4.5 Summary

There should be no doubt that in making morally good clinical decisions, especially those which are challenging or which may be contested by the participants, there must be discussion amongst those on whom the decision impacts, aiming to reach a consensus decision. This chapter argues that, given the plurality and fragmented nature of contemporary society, there can be no absolute universal moral truths. Therefore, in deciding how we should live together, we need to move towards a universally-applicable process-driven moral construct. Compared with the subjective orientation of ethical questions, moral decision making is implicitly aware of the other. As far as possible, all persons affected should be considered. The principles of Habermas' discourse theory of morality, universalizable to all, and communicative action, as a cooperative search for truth, constitute an approach which seeks consensus. Thus, in the clinical encounter, we have an inclusive and non-coercive reflective dialogue.

References

Australian Commission on Safety and Quality in Health Care. 2011. National safety and quality health services standards. In *Australian commission on safety and quality in health care* (ed.). Sydney.

Bertens, H. 1995. *The idea of the postmodern: A history*. London and New York: Routledge.

Candlin, C., Y. Maley, and H. Sutch. 1999. Industrial instability and the discourse of enterprise bargaining. In *Talk, work and institutional order: Discourse in medical, mediation and management settings*, ed. S. Sarangi, and C. Roberts, 323–350. Berlin: Mouton De Gruyter.

Caws, P. 1991. Committees and consensus. *The Journal of Medicine and Philosophy* 16 (4): 375–391.

Crowder, G. 2003. Pluralism, relativism and liberalism in Isaiah Berlin. In Australasian Political Studies Association Conference, University of Tasmania, 29 Sep–1 Oct 2003.

Doran, E., J. Fleming, C. Jordens, C. Stewart, and I. Kerridge. 2015. Part of the fabric and mostly right: An ethnography of ethics in clinical practice. *Medical Journal of Australia* 202 (11): 568–590. doi:10.5694/mja14.00208.

Dworkin, R. 2011. *Justice for hedgehogs*, 1st ed. Cambridge, MA: The Belknap Press of Harvard University.

Fiester, A. 2015. Neglected ends: Clinical ethics consultation and the prospects for closure. *The American Journal of Bioethics* 15 (1): 29–36. doi:10.1080/15265161.2014.974770.

Flyvbjerg, B. 2000. Ideal theory, real rationality: Habermas versus Foucault and Nietzsche. In Political Studies Association's 50th Annual Conference: The Challenges for Democracy in the 21st Century, London School of Economics and Political Science, 10–13 April 2000. London School of Economics and Political Science.

Forst, R. 2014. *Ethics and morals* (J. Flynn, Trans., *The right to justification: Elements of a constructive theory of justice*). New York: Columbia University Press.

Griffioen, S., and R. Van Woudenberg. 1990. We must not forget those who are absent: Interview with Karl-Otto Apel on the universality of ethics. In *What right does ethics have?: Public philosophy in a pluralistic culture*, ed. S. Griffioen. Amsterdam: VU University Press.

Habermas, J. 1972. *Knowledge and human interests* (J.J. Shapiro, Trans.). London: Heinemann Educational.

Habermas, J. 1996. *Between facts and norms* (W. Rehg, Trans., *Studies in contemporary German social thought*). Cambridge: MIT Press.

Habermas, J. 2001a. *Justification and application: Remarks on discourse ethics* (C. Cronin, Trans.). Cambridge: MIT Press.

Habermas, J. 2001b. *On the pragmatics of social interaction: Preliminary studies in the theory of communicative action* (B. Fultner, Trans.). Cambridge, MA: MIT.

Holm, S., P. Gjersoe, G. Grode, O. Hartling, K.E. Ibsen, and H. Marcussen. 1996. Ethical reasoning in mixed nurse-physician groups. *Journal of Medical Ethics* 22 (3): 168–173.

Jennings, B. 1991. Possibilities of consensus: Towards democratic moral discourse. *The Journal of Medicine and Philosophy* 16 (4): 447–463. doi:10.1093/jmp/16.4.447.

Jones, I.R. 2001. Health care decision making and the politics of health. In *Habermas, critical theory, and health*, ed. G. Scambler, 68–85. London: Routledge.

Kerridge, I., M. Lowe, and C. Stewart. 2013. *Ethics and law for the health professions*, 4th ed. Sydney: The Federation Press.

Levinas, E. 1999. *Alterity and transcendence* (M.B. Smith, Trans., *European perspectives*). New York: Columbia University Press.

McDougall, R.J., and L. Gillam. 2014. Doctors' "judgements" and parents' "wishes": Clinical implications in conflict situations. *Medical Journal of Australia* 200 (7). doi:10.5694/mja13.11326.

Outhwaite, W. 1994. *Habermas: A critical introduction (Key contemporary thinkers)*. Stanford, CA: Stanford University Press.

Reichman, N.E., H. Corman, and K. Noonan. 2004. Effects of child health in parents' relationship status. *Demography* 41 (2): 569–584.

Scambler, G. 2001. Introduction: Unfolding themes of an incomplete project. In *Habermas, critical theory, and health*, ed. G. Scambler, 1–24. London: Routledge.

UNESCO. 2007. *Philosophy: A school of freedom* (*Teaching philosophy and learning to philosophize. status and prospects*). In ed. Moufida Goucha, Feriel Ait-ouyahia, Arnaud Drouet, and K. Balalovska. Paris, France: United Nations Educational, Scientific and Cultural Organisation.

Walker, P., and T. Lovat. 2017. Dialogic consensus in medicine—A justification claim. *Journal of Medicine and Philosophy* (In Press).

Walker, P., and T. Lovat. 2016. Dialogic consensus in clinical decision making. *Journal of Bioethical Inquiry* 13 (4): 571–580. doi:10.1007/s11673-016-9743-z.

Walker, P., J. Cassey, and S. O'Callaghan. 2005. Management of antenatally detected lesions liable to obstruct the airway at birth—An evolving paradigm. *International Journal of Pediatric Otolaryngology* 69 (6): 805–809.

White, S.K. 1991. *Political theory and postmodernism* (*Modern European philosophy*). Cambridge: Cambridge University Press.

Wilkinson, D., R. Truog, and J. Savulescu. 2015. In favour of medical dissensus: Why we should agree to disagree about end-of-life decisions. *Bioethics*. doi:10.1111/bioe.12162 (EPub ahead of print April 23).

Wolf, S. 2011. Hiking the range. In *On what matters*, vol. 2, ed. D. Parfit, 33–57. Oxford: Oxford University Press.

Chapter 5
Challenges Facing Dialogic Consensus

5.1 Challenges

First of all, we are aware of criticisms that dialogic consensus may not be practically applicable to the context of political or business philosophy (Flyvbjerg 2000) or to the way administrators or managers enter into discussions. However, we argue that the process of dialogic consensus is applicable to moral philosophy, and especially to decision making in medicine and the health sciences. Critics of dialogic consensus suggest that, in the real world, strategic action in pursuit of self-interest is much more likely to be the dominant approach to situations of dialogue and also in actual decision making. In making clinical decisions, however, we suggest that neither the clinician nor the patient should need to "win" the discussion. The patient has a great self-interest in understanding the situation of illness she is in. The clinician who is aware of empathic compassionate caring and who sees the patient's good as the final purpose of medicine is motivated to understand the concerns and fears of the patient in a way that politicians or lawyers who argue their case in order to change people's ways of thinking, are not so motivated. Furthermore, the dialogic approach which we champion, has potential to meaningfully locate clinical decision making in the concrete realities of the illness at hand, including the nature of the disease itself and its prognosis, as well as contingent circumstances and socio-cultural or religious values.

Second we are going to address issues and thus potential concerns, about how we come to know what we know, power differentials in the dialogue, strategic action versus communicative action, possible moral distress following upon the dialogue, who might be the participants, the importance of time and of the facilitator, and options to manage dissensus.

© The Author(s) 2017
P. Walker and T. Lovat, *Life and Death Decisions in the Clinical Setting*,
SpringerBriefs in Ethics, DOI 10.1007/978-981-10-4301-7_5

5.2 How We Know What We Know

Epistemology is the philosophical term concerned with how we know what we know. That is, it explores the sources, structure and limits of knowledge. *Hermeneutics* is how we interpret knowledge. In Chap. 4, we touched upon Jürgen Habermas' three "ways" of knowing. These are empirical-analytic knowing, historical-hermeneutic knowing, and self-reflective "critical" knowing. Empirical-analytic knowing focuses on the empirical data capture of intrinsic and empirical "facts". Historical-hermeneutic knowing focuses on understanding the meanings of the facts. Self-reflective or "critical" knowing derives from our cognitive interest in discerning truth.

In practice, information about medical diseases and treatments is available from a wide variety of sources. These include the internet, community support groups, magazine and television stories, medically trained friends and relatives, other specialists, and well-meaning but not-otherwise-qualified friends and neighbours. Much of this information may be conflicting and thus confusing to patients. Nonetheless, in complex medical dilemmas, the best way to maximise the patient's Goods, in all their dimensions, may be complex. What information is emphasised and how that information is presented may sway or even hijack the process of dialogic consensus (Walker and Lovat 2016). Clinicians will often begin the dialogue, usually by summarising the medical facts. Those clinicians favouring the medical model will likely present the information from that medical evidence-based perspective. As an example, among ways of "framing" information presented to patients, several factors have been shown to incline the patient towards one or other course of action. These include presenting the percentage chance of *not* developing a particular complication, versus actually developing it, quantifying risks numerically rather than merely listing them, highlighting perceived losses from inaction rather than perceived gains (Aggarwal et al. 2014), amongst many others.

Further, knowledge is inevitably incomplete (Salzman and Lawler 2013). This is because the data-set itself, is incomplete. As well, given their unique past experiences, intellect and character, knowers tend to filter the data they access—that is they tend towards conscious or unconscious selectivity in their hearing of the facts and meanings in the situation at hand. Individuals bring differing socio-cultural perspectives to the way in which they filter the data. As we have already noted, further difficulties with determining knowledge related to normativity include value pluralism, moral pluralism, and short-comings in traditional ethical frameworks. Zygmunt Bauman characterises post-modernity itself as a condition of 'uncertainty' in knowledge (Hugman 2005). There are also practical difficulties in reasoning about complex and medically-challenging situations themselves. Among other considerations, competing principles and consequences, socio-cultural and religious backgrounds and contexts, all impact on the process of coming to know.

However, we argue that these epistemological difficulties notwithstanding, dialogic consensus in medicine remains set upon a firm foundation, in that dialogue collects the facts—of the patient, their medical and family history, their disease, and the socio-cultural context within which it is set. Dialogue then comes to understand the values attached to these facts. For example, again consider a man with prostate cancer. Dialogue may begin with the clinician estimating a percentage for prolongation of five-year survival following radical excision of the prostate and surrounding lymph nodes, versus a less aggressive procedure plus adjuvant radiotherapy. Then the dialogue explores whether, for this patient, risking impotence is worth that prolonged survival percentage. The values of the patient are explored by the parties involved. The dialogue recognises that the values of the patient may not be clearly recognised by him, and the clinician may be asked to guide the patient to understand and make coherent his value structure, and so assist him to make his decisions. The precise meaning of prostate cancer and the risk of impotence will be unique to this patient in his socio-cultural setting. Self-reflective knowing follows when the participants in the dialogue consider the facts and the meanings of the facts, as well as how they impact on the patient and on each other person in the dialogue.

We also argue that motivation to act on a moral decision derives from a process based upon dialogue. A process of dialogue, within a community aware of the principles of discourse theory of morality and communicative action, is based on a philosophy of inter-connectedness (and hence, as we have noted, of intersubjectivity and phenomenology as they apply to the clinical encounter). It is aware of the exposedness and vulnerability which are especially to be found in clinical situations (and a large part of the privilege of being in those consultations as a clinician). Dialogic consensus may not be associated with an empirically factual justification in absolute terms. Recognising the moral basis for the clinician-patient relationship, we argue that in the post-modern era, it is only in the 'moral impulse' (Bauman 1993) of the clinician that morally good decisions will be made—wherein 'each clinician seeks moral responsibility in themselves and looks for it in others' (Hugman 2005).

Impartiality in the dialogue is important. Speaking in the context of politics and political power, Thomas Nagel uses the term *epistemological restraint*—'the distinction between what is needed to justify belief and what is needed to justify the employment of political power depends upon a higher standard of objectivity, which is ethically based' (Nagel 1987). In the context of this book, the fact the speaker holds a belief cannot, of itself, justify its applicability to or acceptance by others. The basis for such restraint is the belief that some things cannot be proven infallibly (while allowing that some things, after investigation and assessment, are a reasonable best belief). The clinician (or anyone else) has no inherent right to force their own beliefs upon others in the dialogue. As well, the deliberations themselves have a moral dimension. The articulated over-arching aim to maximise the Goods of the patient may be the impetus to a positive consensus to an action, or to allow participants to fail to reject an action. In clinical situations, the patient's desires or preferences are often a surreal montage assembled from medical myths, images (Gillett and Amos 2014), and personal desires (as well as those of their relatives). These preferences are but a part of their existence as an ill patient. The patient is entitled to a reasonable medical opinion, and that opinion is subject to critique—so implying legitimacy. Interestingly, clinical consultations using a professional interpreter have been found to be less often associated with reaching the preconditions for communicative action than when a family member is used to interpret (Greenhalgh et al. 2006). Use of a family member as interpreter in clinical consultations is traditionally avoided because of perceived privacy concerns, as well as potential bias or undue influence. Yet there is evidence that the lack of trust, time constraints, and exacerbation of power imbalance identified as being associated with a professional interpreter, are outweighed by the fact that family members are generally trusted by the patient, likely share the lifeworld experience of the patient, and tend to dilute the power differential between clinician and patient.

5.3 Power in the Dialogue

It is necessary that the clinician is aware of what has been termed the epistemology of power (Laura et al. 2008). In this way of knowing, knowledge equates with power. In the clinical consultation there is a power-imbalance between clinicians and patients. This follows upon the near-exclusive access of clinicians to medical knowledge, their higher hierarchical position within the health care system, and perhaps their perceived potential to decline elective treatment. The asymmetry is further encouraged by the importance accorded to quantitative knowledge—traditionally at the expense of meaning and value. Habermas, and Auguste Comte before him, argued that empirical scientific knowledge leads to predictability, and hence to power (Outhwaite 1994). Ron Laura proposes that 'because the epistemic goal of science is to make the world as predictable as possible, the world is stripped of its qualitative dimensions so that only the more predictable quantitative aspects remain' (Laura et al. 2008). Linnie Price notes that medicine's choice to locate itself in science requires that non-observable phenomena be excluded (Price 1984). Laura argues that modern medicine's reductionism recognises only a model of causation, crystallised into what he names 'the "theory of specific etiology"' (Laura and Chapman 2009). This, in turn, has developed historically from Galileo Galilei's, Isaac Newton's and William Harvey's mechanistic world-view which Laura and Chapman characterise as 'fossilised into a metaphysical postulate' (Laura and Chapman 2009), dogmatically favouring an interventionist paradigm in medical decision making. Jeffrey Bishop, in his provocatively titled *The Anticipatory Corpse*, writes that modern medicine is characterised by 'practices aimed at their own practicality ... deploy[ing] a metaphysics of control, of efficient causation ... scientific knowing is an act of power' (Bishop 2011). This follows upon modern medicine's propensity to definitively categorise objects unambiguously, which removes them from further evaluation, and 'is the power to control, to bring about the effects one desires in the world' (Bishop 2011).

This power differential may be encouraged by, for example, the clinicians who are participants in a case conference ignoring or significantly discounting the wishes of the patient or their family (Walker and Lovat 2016). Recall the different language between clinician's judgements and patient's wishes which we discussed in Sect. 4.2. Successfully prioritising communicative rather than strategic action, requires skill and training to minimise the influence of power differentials. How much information the clinician should give depends upon how much information the patient wants, and their level of education and ability to understand. Questions should be determined by empathic connectivity amongst the participants in the dialogue. Power differentials which may impede the ideal speech situation, also exist amongst clinicians, and within families. For example, a junior registrar, who provides day-to-day practical care may not be willing to speak up in the presence of the senior specialist who directs overall treatment for the ward. Some family members may habitually be decision makers for others within the family—for example, the matriarch of an extended family who makes decisions on behalf of a new spouse.

Through self-insight and self-reflection, participants recognise that each member brings to the dialogue their own background, and each needs to be aware of this in themselves and in others. The aim is to establish a situation of non-coercive dialogue, so that a decision about treatment—a consensual decision—can be made. The participants each seek to understand the truths being expressed by each other as the only way to know the good of the patient, in order that it can be maximised. To bring about the best outcome for the patient (and the community) we substitute recognition of values, goodness, and wisdom, rather than emphasising empirical facts and strategic action aimed to coerce or manipulate.

5.4 Strategic Action Versus Communicative Action

Notwithstanding our opening comments to the chapter about dialogic consensus in practical application in the real world of business or politics, at this point it should be acknowledged that a clinician who understands the process of dialogic consensus could deliberately mis-employ it in order to foster strategic action to achieve a desired end-result, rather than communicative action to seek truth. More convincing arguments to persuade the patient in the direction the clinician wants them to go may be more easily articulated by the perhaps more educated and erudite clinician. It is not difficult to recognise when strategic action has come to dominate the clinical consultation. For example, the consultation may begin with the clinician outlining his or her significant expertise in this particular situation, as someone to defer to. Language may be chosen to manipulate the patient's emotions, or by using medical terminology, which the patient does not understand, but which sounds impressive, and fearful, or which appeals to a patient's historical or cultural vulnerability. The paternalistic doctor-patient model may be acting unfettered when a surgeon disallows non-surgical options entirely, via failing to mention them as an option to be considered by the community assembled. Conversely, communicative action would acknowledge all options before exploring their benefits and burdens. In this way, the dialogue seeks to balance the options and their potential consequences for this patient (the approach of proportionism which we favour in the clinical setting—Sect. 3.2), as this patient and their family is helped to reach understanding. In the ideal speech situation, language is chosen which does not aim at manipulating the patient's emotions or responses. Medical terminology will be used, but its meanings will be made clear.

Leaving this aside however, it is likely to be frustrating for the clinician who aims for communicative action that patients from some cultures may believe camomile tea has beneficial effects well beyond those proven by published studies. Beyond factual disputes there is also, increasingly, the likelihood the clinician's moral beliefs are not shared by the community participants. An irreligious clinician may not allow for parents of a multiply-handicapped child, who have a belief in an afterlife as a good place, deciding that going into that afterlife may be a better choice that suffering here on earth.

As a paradigm for moral decision making in medicine, dialogic consensus has a great deal to offer. It may be that, in the clinical setting, a tendency towards strategic rather than communicative action, occurs by way of misunderstanding or ignorance rather than deliberate intent. It would seem an important pre-amble to a case conference utilising this approach, to educate the participants about the process to be followed.

5.5 Moral Distress in the Dialogue

Individual clinicians and patients or their relatives, especially those who seek to be ethically "good", or who habitually seek to make the better decision rather than the worse decision when faced with a choice, may have trouble with what they may see as compromising their integrity in order to achieve consensus. We have just noted that factual knowledge is inevitable incomplete, since data is incomplete, knowers are selective, and individuals bring differing perspectives to the data. As also just noted, Habermas' three "ways" of knowing are empirical data collection and explanation, understanding of meanings and values, and self-reflection seeking truth. While obviously relevant to the mechanics of dialogue in a case conference, the third self-reflective way of knowing is also especially relevant to achieving consensus in the face of a perceived threat to ethical integrity. Since remaining true to one's own principles is lauded in much of Western popular philosophy, literature and film, and "ethical integrity" is viewed as an important tenet by many in society, compromise of one's ethical viewpoint may be associated with considerable tension and lead to distress. Consider, for example, the dialogue about whether to allow clinicians to do more outreach clinics directly to the local indigenous community, rather than require these indigenous patients to make their own way at their own cost, to the potentially culturally unsympathetic big city hospital. This is seen by clinicians as putting patient care first, but administrators can see the potential dilution of their services by spreading their clinicians too thinly, and the cost required to provide additional clinic equipment at locations remote to the fully-equipped existing clinics. Or consider a situation similar to Baby 'H' in Sect. 4.4, but where the mother refuses Caesarean section. She is confident that the team will be able to successfully establish an airway at birth; but if not, accepts this failure, and the death of their child at delivery, as the will of God. Some clinicians will feel inclined to say that they will not be involved—that without Caesarean section they will not risk the death of that child, avoidably in their view, in their operating theatre, with their whole team assembled.

In these situations, where the values of participants in the dialogue differ widely, individuals must carefully and empathically weigh up what is best for the group, despite their own ethical viewpoint (Walker and Lovat 2017, In Press). In order to achieve consensus, each of the participants in the dialogue may need to step away from their most preferred position. That is, they may need to recognise that their own ethical perspective, is not the best perspective with which to view this situation, allowing for the perspectives of other participants in the situation. In the second example above, the clinical team needs to come to understand the value to the family of avoiding Caesarean section, and the family needs to come to understand the value to the clinical team of avoiding the preventable death of a child, while they stand by unable to intervene.

One way to avoid the ethical distress which may follow, is to be aware that conflicting values are not unexpected. Aware also, of the difficulties with the epistemological aspects of dialogue or discourse noted above, we argue that awareness that the dialogue is a moral encounter amongst persons, means that our moral integrity *requires* that we be tolerant of conflicting values. Furthermore, because we are impelled to maintain our commitment to reaching consensus via dialogue, we must remain active in our responsiveness to conflicting values. Importantly, in clinical situations, it also means that the relationship between a doctor and her patient remains on-going despite a value conflict.

Therefore, rather than use the term "conflict of interest", we favour "dualities of interest" as being more appropriate. We all have potential or actual dualities of interest. By this we mean that we all occupy a multiplicity of roles. Often contemporaneous dualities are a fact of life. Tension can occur in certain specific situations where their dualities of interest clash. In these situations (amongst many others), moral conflicts and dilemmas are inevitable. Dialogue seeks to clarify the ethical imperatives in the situation at hand, as each participant perceives them. This lays the groundwork for the resolution of those differences which remain without any sense of ethical capitulation. 'Moral sensitivity' has been characterised as having 'three parts: individual moral worldview, an evaluation of the ethical situation at hand, and the ability to negotiate with others' (Muthusamy 2015). A willingness to negotiate, following upon recognition of responsibility to the other, allows for an openness to re-framing one's perspective, so as to (at least) avoid impasse, and (ideally) be party to a 'joint process of moral learning' (Metselaar et al. 2015). This also brings the discussion back to the framework of proportionism exposited above as our preferred approach in clinical situations.

It is our contention that actual decision making in the clinical encounter should be approached from the actual reality of patients in their situation of illness, set against their socio-cultural background. Hence, it should be approached via a process which seeks consensus after dialogue. Ultimately, if one clinician simply cannot agree with a modality of treatment, then transfer of care to another clinician is an option.

5.6 Participants in the Dialogue

If the dialogue is to be relevant, then those who partake in it—the community that is affected—deserve serious consideration (Walker and Lovat 2016). *Constituency* is the term applied to those who are selected, from the affected community, to take part in the dialogue. It is necessary to clearly articulate who are the participants, and as we noted in the neurosurgical consultation in Sect. 4.3.1, to consider their relative weighting. Increasing the number of participants may or may not increase the likelihood of consensus. Allowing as many members of the community as is reasonable to participate in the process reflects our shared humanity and underlines its importance to the participants and to others who may not be participating directly, but merely observing it. When considering decisions for child members of a family, the age and maturity of siblings, for example, will help determine whether they should be included, formally or informally, or not at all. *Stakeholder analysis* is a tool which has as its primary objective 'to map the power, interest and influence of relevant stakeholders around a decision' (Kerridge et al. 2013). Stakeholders are those who have an interest in the outcome of a problem. To the extent that professional and regulatory bodies have an interest in the outcome, members of the dialogue need to be aware of the tenets of their professional codes of ethics and similar statements, as part of the starting point for the dialogue. Beyond questions about constituency, the analysis estimates the salience, relevance, or significance of each stakeholder by characterising their dynamic of power, legitimacy, and urgency (Montgomery and Little 2001).

5.7 Time and the Facilitator

In clinical moral decision making, taking the required time to engage in the dialogue as fully as possible has already been mentioned as desirable, but is also a necessary constraint (Walker and Lovat 2016). That is, it can take time for the patient or their relatives to come to a full understanding of the medical condition and what its prognosis might be. More than one dialogue may need to be scheduled. Eventually, however, a decision needs to be made from among possible treatment options. There may be changes in the clinical situation over time, requiring re-evaluation and so prompting a further dialogue within the community. In clinical scenarios, there may well be a time-constraint on the decision-making process itself, if not of hours, then conceivably of days. Moral philosophical validity also entails an important dimension of time—in the sense of initiating a process of ethical reflection over time, rather than making a discrete and self-contained moral decision at a single point in time. Habermas' discourse theory of morality requires that participants accept the consequences which can be anticipated. This may mean that the morally good decision will change during the process of on-going dialogue and as the clinical situation evolves.

Pre case-conference dialogic discussion may be a means for providing factual information about the clinical condition. This may also be useful when opinions seem to vary widely—as a way of encouraging participants to consider the perspective of the other, and perhaps have the opportunity, after self-reflection and consideration of the patient's actual situation in the world, to moderate an extreme stance.

The facilitator of a case conference has a pivotal role to play in achieving the practical realisation of the requirements for meaningful dialogue. The most appropriate facilitator of the dialogue may be a senior member of the nursing staff or an allied health worker, or a psychologist, social worker, or chaplain. These professionals may have more skills in facilitating dialogue than the clinician—though they may not. A facilitator who is not the patient's usual clinician may be less-threatening to the values of the participants, perhaps because they may not themselves be categorised as the interventionist, that is for example, the one who will be withdrawing treatment for their loved one if that is the consensus decision, and who thus may be perceived as having something of a vested interest in a particular outcome. The patient herself, although the principal actor in the dialogue, unless she has very special training, may be too close to the outcome to be able to effectively facilitate it. Regardless, however, a self-insightful facilitator will welcome shared facilitation if it develops during the dialogue.

5.8   Dissensus

Where the dialogue fails to reach consensus there remain a number of options (Walker and Lovat 2016). These include (Beauchamp 2009): (1) Specification. This is the approach of narrowing a general norm to the specific context at hand. General norms may be too broad to be useful. Consider the general norm "do whatever it takes to keep the patient alive", to which a family may hold steadfastly. Specification narrows the general norm to the specific context of the particular patient with the particular diagnosis and prognosis in her particular historic-cultural background, in this ICU. Granted this specific context, the family may come to understand that their mother's good is best served by allowing her to die, in order for example to end her suffering. (2) Gatekeeping. When, for example, resources are very limited, adopting a gatekeeping type of policy may be helpful. One example is that Intensive Care admission be restricted to those under a certain age. (3) Collection of the facts. Obtaining factual information about any apparent disagreement may show that it is not a moral dilemma at all. (4) Definitional clarity.

Providing definitional clarity means exploring what each party understands as the meaning of the language which is being used in the dialogue. Additionally, using examples and counter-examples to contextualise this particular dilemmatic situation, may be useful, and may identify the reasons the participants hold the beliefs they hold. While they may be reasonable, they may in fact be culturally-based, based upon ignorance, or in fact be illogical and understanding their incoherence may result in its resolution and hence consensus. Another response is to make it explicitly clear that the clinician is not in an authority figure role—which is a central trigger for some personalities, for example.

Where dissensus remains, another option is mediation. For example, referral to a clinical ethics committee (to distinguish it from a research ethics committee). This committee seeks to rebuild consensus 'by taking as many different perspectives into account as possible and, through an iterative process of consultation, ensuring that all parties agree … In practice it provides an effective means of mediating between widely divergent perspectives to generate agreement' (Kerridge et al. 2013).

It remains to be said, however, that there are at least four situations in which participants may exclude themselves from the dialogue (Walker and Lovat 2016).

The first situation is when serious disagreements are recognised, not in the sense that they follow complex multi-factorial arguments, but in the sense that, what we will call "the world view" (that is, the entire approach to life) of one or more of the participants in the dialogue, are so far apart that it would require an impossibly long period of argument to find the common ground needed to resolve the disagreement that might not, even then, ever be possible. An example would be in the argument that it is wrong to interfere with the natural history of disease and death, by preventing it—because it is God's will that disease and death occur. This belief is not possible to oppose by recourse to logic, or to any possible medical trial design. To generalise further, communicative action is only conceivable against a background of broad agreement concerning at least the basic features to be submitted to argumentation—that is, 'it is impossible to problematize all factual or normative claims simultaneously' (Cronin 2001).

The second situation is when there exists an abnormal psychology or pattern of faulty thinking, in one or more of the participants in the dialogue, be they in the clinician, or the patient or a relative. An example would be a participant who has no or little empathic awareness of the perspective of others (for example, a sociopath). This will likely de-rail the process. In this situation, it is not unreasonable to set limits to the obligation to respond to arguments. Members of the dialogue with cognitive impairment are a similar challenge. If affected by the outcome of the dialogue, then their contribution needs to be imputed wherever possible.

The third situation is when patients or their surrogates adopt either of the extreme stances of "only the patient/surrogate can decide" or "only the doctor can decide". If they cannot be educated that they are entering into a mutually-consensual dialogue, then they have chosen to exclude themselves from the process of a meaningful dialogue. This situation is able to be recognized in certain cultural groups. In these groups, the voice of a dominant single person, for example, a matriarch-figure is deferred-to by other members who have been invited to be part of a mutual dialogue, but who are reticent to oppose the dominant person. While it may be possible to educate the participants that each member of the dialogue has an equal co-responsibility to contribute to the consensual decision, there might not be any willingness on the part of the participants to engage.

The fourth situation is when members of the community entering into dialogue "already know the answer" and are simply unwilling to listen to the other, or to be open to the possibility of moving from their preferred ethical position.

Finally, psychological factors are, empirically, very well-recognised overlays which influence decision-makers, generally in ways of which we are completely unaware (Appiah 2009). Some of these factors include reacting more compassionately in less stressed, quieter or even more 'pleasantly-smelling' environments. Additional potentially confounding factors include our personal socio-cultural and historical filters developed as a result of the influence of our parents, our formal and informal schooling, religious ideations, and influence of the media. We have noted above in Sect. 5.2 that accessing knowledge involves a certain selectivity, and this may extend to the application of different ethical standards to ourselves, compared with those we apply to others. A particular factor is that which may be termed "tribal loyalty", which can act as a strong moral motivator. Perhaps able to be explained as an evolutionary imperative, tribal loyalty means that we tend to protect and to favour in our decisions, those in our own tribe. Conceivably, we can generalise from our understandable loyalty to our children, to our family, to other people close to us personally or professionally to the wider group of "people-like-us". This means that those who are different from us—in terms of geographical origin, physical appearance, or in their cultural, religious, social or moral values—may not be persons we tend to protect and favour. In fact, this sense of difference might well lessen our proclivity to see their values in a favourable moral light. We may make moral decisions influenced by these tribal loyalty factors, rather than make a genuine attempt to understand where they are coming from.

One crucial and practical way to avoid this is to have a dialogue, amongst all those affected, so as to come to recognise and minimise these psychological motivators.

5.9 Summary

This book recognises that even with theoretical difficulties placed to one side, there are complex and emotionally-coloured processes at work in the dialogue consequent upon the humanity of the participants. These are further liable to engender an amount of tension within the process itself. The potential tension is greater, along with an equivalent potential to impel strategic action, when the diversity of personal or socio-cultural beliefs of the participants are greater. Striving to understand and adopt the perspective of others in the dialogue facilitates the process of mutual cooperation.

Nor do we deny the practical difficulties associated with the process of dialogic consensus. It aims to achieve the best possible decision among human beings discoursing under inevitably emotion-laden circumstances. We believe that, practical difficulties notwithstanding, the process of dialogue and consensual decision making impels a motivation to act upon the decision. Closely allied to this, it also gives participants *permission* to act upon the decision. Consider the situation where, in order to save a life, a potentially heroic, expensive or risky intervention is required. The decision reached by the community of those affected is, on balance, not to proceed with the intervention. Aware of the normative moral *oughtness* or *shouldness* inherent in the decision made this way, all the members of the community involved can together look towards palliation and allow life to end. As we have noted in Chap. 4, this has potential to be highly reassuring to family members who have been involved in the decision.

References

Aggarwal, A., J. Davies, and R. Sullivan. 2014. "Nudge" in the clinical consultation—An acceptable form of medical paternalism? *BMC Medical Ethics* 15 (1): 31. doi:10.1186/1472-6939-15-31.

Appiah, K.A. 2009. *Experiments in ethics. Mary Flexner lectures at Mryn Mawr College*, vol. 3. Harvard University Press.

Bauman, Z. 1993. *Postmodern ethics*, 1st ed. Oxford: Wiley-Blackwell.

Beauchamp, T.L. 2009. Moral foundations. In *Ethics and epidemiology*, 2nd ed, ed. S.S. Coughlin, T.L. Beauchamp, and D.L. Weed, 39–41. Oxford: Oxford University Press.

Bishop, J.P. 2011. *The anticipatory corpse: Medicine, power, and the care of the dying*. Notre Dame, IN: University of Notre Dame.

Cronin, C. 2001. Translator's introduction (C. Cronin, Trans.). In *Justification and application: Remarks on discourse ethics*, xi–xxxi. Cambridge: MIT Press.

Flyvbjerg, B. 2000. Ideal theory, real rationality: Habermas versus Foucault and Nietzsche. In *Political Studies Association's 50th Annual Conference: The Challenges for Democracy in the 21st Century*, London School of Economics and Political Science, 10–13 April 2000. London School of Economics and Political Science.

Gillett, G., and C. Amos. 2014. Words are not just things. In The New Zealand Bioethics Conference, Dunedin, Otago, 26 January 2014.

Greenhalgh, T., N. Robb, and G. Scambler. 2006. Communicative and strategic action in interpreted consultations in primary health care: A Habermasian perspective. *Social Science and Medicine* 63 (5): 1170–1187. doi:10.1016/j.socscimed.2006.03.033.

Hugman, R. 2005. *New approaches in ethics for the caring professions*. Hampshire: Palgrave MacMillan.

Kerridge, I., M. Lowe, and C. Stewart. 2013. *Ethics and law for the health professions*, 4th ed. Sydney: The Federation Press.

Laura, R.S., and A. Chapman. 2009. *The paradigm shift in health*, xiii. Lanham: University Press of America.

Laura, R.S., T. Marchant, and S.R. Smith. 2008. *The new social disease: From high tech depersonalisation to survival of the soul*. New York: University Press of America.

Metselaar, S., B. Molewijk, and G. Widdershoven. 2015. Beyond recommendation and mediation: Moral case deliberation as moral learning in dialogue. *American Journal of Bioethics* 15 (1): 50–51. doi:10.1080/15265161.2014.975381.

Montgomery, K., and J.M. Little. 2001. Ethical thinking and stakeholders. *The Medical Journal of Australia* 174 (8): 405–406.

Muthusamy, A. 2015. Shared language and moral sensibility in resolving clinical ethics conflicts. *The American Journal of Bioethics* 15 (1): 60–61. doi:10.1080/15265161.2015.975582.

Nagel, T. 1987. Moral conflict and political legitimacy. *Philosophy & Public Affairs* 16 (3): 215–240.

Outhwaite, W. 1994. Scientism in theory and practice. In *Habermas: A critical introduction. Key contemporary thinkers*, 35. Stanford, CA: Stanford University Press.

Price, L. 1984. Art, science, faith and medicine: The implications of the placebo effect. *Sociology of Health & Illness* 6 (1): 69. doi:10.1111/1467-9566.ep10777362.

Salzman, T.A., and M.G. Lawler. 2013. Method and Catholic theological ethics in the twenty-first century. *Theological Studies* 74: 903–933.

Walker, P., and T. Lovat. 2016. Dialogic consensus in clinical decision making. *Journal of Bioethical Inquiry* 13 (4): 571–580. doi:10.1007/s11673-016-9743-z.

Walker, P., and T. Lovat, T. 2017. Dialogic consensus in medicine—A justification claim. *Journal of Medicine and Philosophy* (In Press).

Chapter 6
Conclusion: Looking to the Future of Moral Decision Making in Clinical Settings

6.1 Looking to the Future

If there is a single rationale for a book such as ours, it is that there is a need, in our contemporary era, for a moral philosophical framework which contains principles of conduct towards other persons. Predicated upon the notions of interconnectedness, intersubjectivity, and alterity, this should recognise that principles of conduct towards others can, *and must*, be determined, no matter how one's own ethical values, conceptions of the good, or life-choices differ from those of others.

Thus, we have focused on moving forward philosophically from the observation that as clinicians trying to make morally good decisions with our patients, we need to have a discussion, a conversation, or a dialogue with our patients about their care. We have aimed to place the need and mechanism for this dialogue onto a more robust moral philosophical footing than is currently commonplace. Such an approach must be orientated towards a responsibility to others, and so necessarily be open, in the clinical encounter, to the difference, dissonance and ambiguity we alluded to in Sect. 4.1.

In summary, we have argued that we need to move away from traditional appeals to substantive ethical frameworks or to principles derived from them, which constitute a relativistic and essentially subjective ethical construct. Given the plurality and fragmented nature of contemporary society, and the recognition of alternative viewpoints, in order to make morally good decisions we need to articulate a process-driven method for making those decisions. We need to move towards an inclusive, non-coercive reflective dialogue in order to ensure a more universally-applicable moral construct. Jürgen Habermas' discourse theory of morality generalises the Kantian categorical imperative as determined by ethical monologue, to a wider consensus-seeking dialogue. Via practical discourse amongst the participants, inter-subjective contextual interpretation is incorporated via a process of reflective dialogue. Thus consensual agreement is reached about what constitutes morally-correct action through an integrative balance between

© The Author(s) 2017
P. Walker and T. Lovat, *Life and Death Decisions in the Clinical Setting*,
SpringerBriefs in Ethics, DOI 10.1007/978-981-10-4301-7_6

deontology and teleology. The process we argue for, seeks dialogic consensus, is implicitly both cognisant of the other and contextualised to the particular situation at hand, and considers all persons affected by the decision. Intersubjective consensus imbues the decision with normative force. This in turn renders the process action-guiding.

In order to justify this approach, we have explored two areas. First, we considered how we actually come to know what we know (*epistemology*). We have discussed Habermas' cognitive "ways" of knowing, and also what we can understand from the lived experience of phenomenology. We submit that the dialogue between clinician and patient aims to collect the facts of the patient's clinical situation of illness, together with the amount of family support and their wider socio-religious and cultural setting, and then the dialogue seeks to understand the values which the patient attaches to these facts. Second, we consider the action-guiding sense of oughtness or shouldness attached to a consensual decision (*normativity*). This is based upon Habermas' discourse theory of morality (universalizable to all, all of whom accept the consequences) and the principles of communicative action (a cooperative search for truth). Our approach is predicated upon the awareness that, especially in our current era, moral conduct towards others must be able to be determined despite the different ethical values we hold as patients and as clinicians.

Thus, from a philosophical perspective, the moral discourse is relocated away from a first-person monologue or a third-person abstraction, into the second-person perspective—because we are aware of the intersubjectivity inherent in our relationships with our patients. It is in this space where the ultimate reference point is in the 'sociality or connectedness of people' (Verlinden 2010). Thus, we recognise the importance of moral decisions reached via a process of active and reflective communicative consensus. We believe that the decision we come to via dialogic consensus has both epistemic and normative force.

Moral philosophy is a very personal but very serious enquiry into how we should behave, how we should treat other humans, and what inner motivations best enhance our human dignity. As well as of those others with whom we co-exist and are necessarily in intersubjective relationships. In clinical contexts, this especially applies to our patients and the privilege we are afforded in being able to be in a caring relationship with them. In our daily interactions with patients, clinicians deal with intensely personal and often serious situations. This book holds that it is perfectly reasonable to expect that clinicians approach moral decision making situations with appropriate gravitas. It is founded upon the premise that clinical interactions have a necessary moral philosophical dimension. This holistic approach suggests that part of the task of clinicians is, as a corollary, to support the wider moral community in its development.

As Ludwig Wittgenstein asserted, 'philosophy unties knots in our thinking. Hence its results should be simple. But philosophising has to be as complicated as the knot it unties' (Wittgenstein 2011, [Prop 21]). As medical educators, our responsibilities are even greater. Intellectual knowledge alone is insufficient—'the teacher has to be able to stand behind his words, to be present in his words' (Cowley 2011), paraphrasing Gaita (1991, 2004).

This book has sought to interweave the threads of clinical medicine with those of moral philosophy. Edmund Pellegrino (having delineated the four Goods of the patient), articulated that:

> Medicine is at heart a moral enterprise and those who practice it are de facto members of a moral community. We can accept or repudiate that fact, but we cannot ignore it or absolve ourselves of the moral consequences of our choice (Pellegrino 1990).

As has been argued in this book, discourse and dialogue are predicated upon language. It is language which makes the moral concept transferable. The seeking of wisdom, as characteristic of moral philosophy (Sect. 1.1), and re-visited in the virtue ethics framework (Sect. 2.5), is a necessary part of being a good clinician. It is wisdom not in the sense merely of a wise person saying clever things, or being able to make judgements about the actions of others. Rather, it is wisdom in the more meaningful sense that 'in his presence and under his compassionate attention one is oneself more inclined to say wise things'—pointing to the importance of wisdom in achieving meaningful dialogue (Cowley 2011). Wisdom, in the view of Rhett Gayle, may be characterised by four qualities: (1) an ability to discern appearance from reality; (2) an awareness of the limits of our own knowledge and understanding; (3) a desire for a good outcome; and (4) self-knowledge, with 'ongoing transformative impact on the knower', described by him as the core of 'choosing well' (Gayle 2011). This wisdom, this understanding, rather than merely knowing, segues into values, and hence morality, and is integral to decision making in clinical contexts.

As we have shown, once we recognize that in making moral decisions, appeals relevant to earlier epochs—be that to gods, to God, or to rationalism—are no longer apposite, the basis for moral decision making must look to a process, rather than to a substantive normative or theistic framework alone. To achieve normative force, a process of discourse communication must follow, wherein the participants in the dialogue are drawn from the community affected and follow principles of communicative action. That is, intersubjective consensus after dialogue. We do not deny the practical difficulties associated with the process, but we believe that, practical difficulties notwithstanding, the process of dialogue and consensual decision making has very important implications for clinical interactions.

References

Cowley, C. 2011. Moral philosophy and the 'real world'. *Analytic Teaching and Philosophical Praxis* 31 (1): 21–30.
Gaita, R. 1991, 2004. *Good and evil: An absolute conception*, 2nd ed. London: Routledge.
Gayle, R. 2011. Befriending wisdom. *Analytic Teaching and Philosophical Praxis* 31 (1).

Pellegrino, E. 1990. The medical profession as a moral community. *Bulletin of the New York Academy of Medicine* 66 (3): 221–232.

Verlinden, A. 2010. Reconciling global duties with special responsibilities: Towards dialogical ethics. In *Questioning cosmopolitanism. Studies in global justice*, ed. S.V. Hooft, and W. Vandekerckhove, 83–104. Dordrecht, NL: Springer.

Wittgenstein, L. 2011. A Wittgenstein primer. In ed. T. Lowes. Mobi Books.